BALIDAN

BALIDAN

STORIES OF INDIA'S GREATEST

PARA SPECIAL FORCES

OPERATIVES

SWAPNIL PANDEY

HarperCollins *Publishers* India

First published in India by HarperCollins *Publishers* 2023
An Imprint HarperCollins *Publishers* India
HarperCollins *Publishers* India, Cyber City, Building 10-A,
Gurugram, Haryana – 122002, India
www.harpercollins.co.in

This edition published in India by HarperCollins *Publishers* 2025

10

P-ISBN: 978-93-5699-355-6
E-ISBN: 978-93-5699-356-3

Typeset in 11.5/15.2 Adobe Garamond at
Manipal Technologies Limited, Manipal

Printed and bound at
Repro India Limited

MIX
Paper | Supporting
responsible forestry
FSC® C047271

This book is produced from independently certified FSC® paper
to ensure responsible forest management.

HarperCollins *Publishers*, Macken House, 39/40 Mayor Street Upper, Dublin 1,
D01 C9W8, Ireland

शौर्यम ... दक्षम ... युध्धेय...! बलिदान परम धर्म!
(वीरता के युद्ध में, बलिदान सर्वोच्च धर्म है)

*For all the nameless, faceless, fierce and worthy owners of the
Balidan badge; the legendary Para Special Forces operatives who have
given their all to the nation, and to their families who have stood by
these real-life superheroes in times of war and in times of peace.*

What manner of men are these who wear the maroon beret?

THEY ARE, FIRSTLY, all volunteers, and are toughened by hard physical training. As a result they have that infectious optimism and that offensive eagerness which comes from physical well being.

They have jumped from the air and, by doing so, have conquered fear.

Their duty lies in the van of battle: they are proud of this honour and have never failed in any task.

They have the highest standards in all things, whether it is a skill in battle or smartness in the execution of all peace time duties. They have shown themselves to be as tenacious and determined in defense as they are courageous in attack.

They are, in fact, men apart—every man an Emperor.

Field Marshal Bernard Law Montgomery,
1st Viscount Montgomery of Alamein

The Parachute Regiment and Para Special Forces

THE PARACHUTE REGIMENT of the Indian Army is primarily an airborne regiment, specializing in parachute operations and quick deployment behind enemy lines. It is an arm of the infantry forces. It traces its origin to the spectacular success of German paratroopers during the initial battles of the Second World War, which caught the attention of all military thinkers. Gen Sir Robert Cassels, the commander-in-chief of the Indian Armed forces, ordered the raising of an Airborne Cadre in October 1940. The initial battalions consisted of the 50th Parachute Brigade raised at Delhi Cantonment in 1941 with Brig W.H.G. Gough as the first Commander. The first Indian officer to join the Parachute Brigade was Lt Rangaraj, the medical officer of the 15th battalion of the 50th Parachute Brigade. The Indian Parachute regiment was raised on 1 March 1945 with Lt Gen Frederick (Boy) Browning as the colonel of the regiment. The regimental insignia was that of its counterpart in the British Army, which was worn on a maroon beret. After the Second World War the Indian Parachute regiment was disbanded in 1945 and placed under the 2nd airborne division consisting of three parachute brigades, i.e., the 50th, 77th and 14th airlanding brigades, as part of the demobilization of much of the Indian Army. After Partition, the 50th and 77th remained in India, which were redesignated as the 1st, 2nd and 3rd battalions of the Para Regiment when the Indian Parachute Regiment (Para) was raised on 15 April 1952.

Thereafter, 4 Para was raised at Agra in 1961, and after the 1962 war, the regiment was expanded by raising five new Parachute battalions. After the 1965 war 9 Para was raised on 1 July 1966 to take on commando-type operations. A year later 9 Para was split to form 10 Para. In 1969 the titles of both units were suffixed with the word 'Commando'; this was changed to 'Special Forces' in 1994. The Indian Maroon fraternity also comprises Combat Support Arms and Service units of the Parachute Brigade, the only one of its kind in the Indian Army. Various territorial armies, Army Service Corps (ASC), field regiments, etc., are part of it. The Parachute regiment got its colours from President Zakir Husain on 6 October 1967.

The motto of the Parachute Regiment is 'Shatrujeet'. The peculiar badge was designed by Capt M.L. Tuli of 2 Para. It is inspired by King Shatrujeet of Hindu mythology. The formation sign of the Para Brigade represents a mighty lightning bolt from the sky and the triumph of good over evil. 'Balidan', the badge of sacrifice, signifies the unconventional warfare missions of the Special Forces, and is worn by the personnel of Special Forces of the Parachute Regiment, i.e., Para (SF). The origin of the force goes back to a special task group called Meghdoot Force, raised by Major Megh Singh on 1 September 1965. 'Special Forces' is a term used to describe relatively small military units raised and trained for reconnaissance, unconventional warfare and special operations. These exclusive units rely on stealth, speed, self-reliance and close teamwork, as well as highly specialized equipment. Traditionally, the mission of the SF are in five areas: counterterrorism, unconventional warfare, facilitating the internal defence of foreign countries, special reconnaissance and direct action against specific targets.

Contents

List of Acronyms

AC	Ashok Chakra
KC	Kirti Chakra
SC	Shaurya Chakra
SM	Sena Medal
VSM	Vishisht Seva medal
CO	Commanding Officer
CRPF	Central Reserve Police Force
IMA	Indian Military Academy
JCO	Junior Commissioned Officer
LAC	Line of Actual Control
LeT	Lashkar-e-Taiba
LI	Light Infantry
LoC	Line of Control
NDA	National Defence Academy

OGWs	Over Ground Workers
Para (SF)	The Parachute Regiment's Special Forces units/personnel
PLA	People's Liberation Army of Manipur
PoK	Pakistan-occupied Kashmir
PT	Physical training
QRT	Quick Response Team
RR	Rashtriya Rifles
SADO	Search-and-destroy operation
SF	Special Forces
SM	Subedar Major
SSB	Services Selection Board
UAV	Unmanned aerial vehicle

Author's Note

THIS BOOK IS the product of more than 200 interviews conducted over a one-year period, multiple trips to various Special Forces (SF) units, seeking permission from the Indian Army to interview SF personnel all over the country and correspondence with sources in India and elsewhere. The hair-raising tales of unimaginable bravery against all odds sets the SF a class apart. While I was putting this collection of stories together, words often fell short. The legacy of this league of warriors, the righteous owners of the title of 'Balidan'—the Indian Para (SF)—is shrouded in mystery and is the stuff of legends. Why do I say this? Because if you try to reach out to one of these men, you'll discover that behind their popular names in public—the Fourth of North, Ghosts, Vipers, Red Devils, Predators, Desert Scorpios—they don't really exist.

Writing this book has been a kind of spiritual awakening for me, opening doors of self-realization that eventually led to much self-improvement in my life. Sitting with these passionate young men, who have achieved and done so much for our nation, listening to their stories in their own words—it has changed me as a person. I started writing *Balidan* in January 2021 and finished in January 2023. What an incredible journey it has been!

I was privileged to witness these men jumping from their aircraft and combat-diving in real time. Over the course of my research for this book, I also met operatives who have crossed enemy borders and survived treacherous bouts of fighting in the mountains or in extremely adverse weather conditions. The high-risk adventure of extreme sports from combat at sea and skydiving to mountaineering have also been integral part of their job profile. Their exceptional physical and mental capabilities are backed by a unique regimental upbringing and rare military skills. Yet, it is not merely their training or skill behind enemy lines that makes these men stand out in a crowd. It is their ability to detach themselves—under pressure—from human emotions like love, fear, anger or excitement that lends them an indefinable air of spiritual toughness.

There is a reason we call them special.

This book is a miracle. I never thought I would be able to complete it—in light of the sheer scale of the hitherto-unheard and unimaginable stories I was privy to or the many difficulties I faced as a storyteller in the process of meeting these extraordinary men and being allowed to write their extraordinary stories. It is indeed a rare privilege. In the era of modern technology, information hybrid warfare and global threats, when the world is going through changes, it is important we provide our next generations with national icons of heroism to inculcate feelings of patriotism and instil confidence in our nation's abilities.

These men can be compared to Greek mythological heroes like Hercules, Achilles or Odysseus. They often enjoying a cult following among men. Fervour and utter professionalism have been the trademarks of these extraordinary men while they uphold the zenith of valour and steadfast devotion to duty, making hundreds of sacrifices along the way.

I would like to mention that while the professional achievements of these men are spectacular, their off-duty escapades are also most colourful. We must not forget that we are talking about rebels from the Armed Forces, the antithesis to military culture and discipline.

Their romances are the stuff of legendary love stories, and when they fight, battles are born! Whether it is the 21 Para (SF), the 9 Para (SF) or the 4 Para (SF), they are all legendary battalions from the remarkable world of Indian Para (SF), which has survived several definitive epochs in the history of the Special Forces. Many a time, I have been anxious about my abilities to do justice to these stories. The Indian Army is notoriously protective of the privacy of these men and their missions—getting permission from the requisite authorities has been a special operation in itself. Despite the Army's generosity, there are still many stories that had to be left out of these pages for reasons of security and privacy.

Yet another focus of this book is the role that the families—wives, girlfriends, children and ailing or elderly parents—of the Special Forces play. They have to learn to live without the men they love—and in this lies a special kind of heroism, a courage that is not really talked about or celebrated. I wanted to illuminate this silent bravery that allows the family to understand and empathize with the missions and careers that their men have chosen.

The rare privilege of writing on legends like Brigadier Saurabh Singh Shekhawat, KC, SC, SM, VSM, or Subedar Major/Honorary Captain Mahendra Singh, KC, SM, was unbelievable. To meet them and listen to their mythical stories in their own words was the opportunity of a lifetime, a storyteller's dream. The fact that they had survived various eras in the history of the Special Forces—from the 1990s through the peak of terrorism to the current times, when modern warfare is threatening the safety of the nation—and achieved so many milestones along the way, which in turn inspired entire generations of Special Forces operatives, is astounding.

Then there are young operatives like Capt Tushar Mahajan, SC (posthumous), or Paratrooper Chhatrapal Singh, SM (posthumous), who are the dashing young superstars of the Special Forces. They looked, behaved and dressed like kings and were killed in action at a young age. Listening to their comrades-in-arms relating a hundred fascinating anecdotes about them was again an otherworldly experience.

This is the reason why I feel it is terribly unfortunate that while Hollywood has produced hundreds of movies on the exploits of the United States Navy Seals or the Special Air Service (SAS) of the British Army, soft-propagating their reverence all across the world, there are hardly any movies on the Indian Special Forces when they are the best in the world. I hope this book plugs that void.

At the end of the day, *Balidan: Stories of India's Greatest Special Forces Operatives* is a book that makes me proud, even though I have only been able to tell six stories. Each story is different because of the men who inhabit them; each is unique because it is representative of the thousands of untold stories from the covert world of the Special Forces. I hope this book evokes respect and gratitude in the hearts of my fellow citizens. Over the course of my research, I met many Para (SF) operatives who confessed that they had joined the Special Forces after listening to stories of legends like Capt Arun Singh Jasrotia or Colonel Santosh Mahadik. I hope this book provides inspiration to India's youth to join the SF, pursue extreme sports or will at least inspire them to challenge their limits and achieve all their dreams.

In sum, this is a book dedicated to the real superheroes of our nation. It could not have been written without their invaluable assistance.

Jai Hind!

21 Para Special Forces: The Waghnakhs

THE 21 PARA Special Forces are also popularly known as the Waghnakhs (Tiger's Claws). The name comes from a deadly Maratha weapon dating back to the era of Chhatrapati Shivaji. The 'Bagh-Nakh'[1] consisted of four or five curved blades affixed to a crossbar or glove designed to slash through skin and muscle. It is believed to have been inspired by the armature of a tiger. The weapon was used by Shivaji to heroically kill the military leader of Bijapur Sultanate, Afzal Khan.

Today, the Waghnakhs are the sentinels of India's Northeast. The battalion was raised as part of the Baroda State Forces as the 2nd Baroda Infantry in the 1860s and was later amalgamated into the Indian Army as the 21 Maratha Light Infantry (LI). Then, the battalion was dismantled and merged into the 20 Maratha LI, which later became the 10 Mechanised Infantry. On 11 February 1985, following Operation Bluestar, 21 Maratha LI was reraised at Belgaum. After its successful tenure in Operation Rhino, it was

---

1 During those times, the weapon was called 'Bagh-Nakh' or 'Wagh-Nakh'.

selected for conversion into a Para (SF) battalion. On 1 February 1996, after meeting the stringent probation requirements, it was rechristened as 21 Para (SF), thus becoming the only Para (SF) battalion to be converted from an existing infantry battalion.

The 21 Para (SF) specializes in jungle warfare, operating in rural and urban scenarios; they are famed for their lethal abilities to carry out cross-border surgical strikes, such as the ones they carried out in the border areas of Myanmar in retaliation to an ambush on an Army convoy in 2015.[2] The success of this particular operation can be judged from the fact that not one casualty was reported on the Army's side; in fact, the Waghnakhs destroyed two terrorist camps inside Myanmar, along the Nagaland and Manipur borders, at two different locations. This battalion also has the unique distinction of having served in all four sectors of Jammu and Kashmir—Drass, Mushkoh, Kaksar and Batalik—during the Kargil War.

The 21 Para (SF) has an outstanding record of having received three Kirti Chakras, sixteen Shaurya Chakras, one Bar to Shaurya Chakra, two Yudh Seva Medals, three Vishisht Seva Medals, fifty-four Sena Medals, seventy-five Chief of the Army Staff Commendation Cards, one Mention-in-Despatches[3] and numerous other gallantry awards, reflecting its mettle as a counterinsurgency specialist battalion. The unit's role in Operation Hifazat, Operation Rhino and others has been instrumental in strengthening national security, especially in the Northeast.

---

2   https://www.news18.com/news/india/army-commandos-of-21-para-entered-myanmar-to-neutralise-northeast-terrorists-1004258.html
3   Based on data available till 2020.

1

Colonel Santosh Yashwant Mahadik, SM: SC (posthumous): The Waghnakh Who Lived and Died for the Nation

1988
Sainik School, Satara
Maharashtra

A group of twelve-year-old boys stood before the Shivling in the Mahadev Mandir at Sainik School, Satara. The boys were on a mission, and Santosh was their leader. The legend of Chhatrapati Shivaji Maharaj so inspired Santosh that he cut his little finger and let the blood drip onto the shivling. Under his breath, he swore an oath to serve the nation till the day he died. In doing so, Santosh was subconsciously following in the footsteps of Chhatrapati Shivaji himself, who took an oath of 'Hindavi swarajya' at the Raireshwar temple near Pune in 1645.

SANTOSH YASHWANT MAHADIK was born into a humble family. His father and brothers were dairy farmers who supplied milk door to door in their little village called Pogarwadi in Satara district.

Santosh's father, Madhurkar Ramchandra Ghorpade, was associated with social service in the village, with local residents often turning up to seek his advice or help. The large-hearted man that he was, Ghorpade rarely turned down anyone's request, always striving to selflessly serve his community.

None of this was lost on young Santosh, who had dreamt of becoming a soldier since the age of five. Patriotism ran deep in his family's blood, as did a desire to give Santosh a better life even though they didn't have much. His father eventually enrolled him in the residential Sainik School in Satara, where he began studying in Class Six. The rest, as they say, is history.

Santosh's childhood was spent immersed in deep-rooted values, tradition, culture and familial affection. With Ghorpade busy with his business and social service, and Kalinda, Santosh's mother, was involved in looking after their joint family, Santosh was left in the care of Kalinda's parents, Yashwant Bala Mahadik and Babai Yashwant Mahadik,[1] who lived with their daughter and son-in-law in the same house. Santosh's elder brother, Tatya, also took on the responsibility of caring for him. It was Tatya who attended the school's parent–teacher meetings, bought his schoolbooks and mended his torn clothes.

As his friends recall, Col Santosh Mahadik was not only academically brilliant but also an ace football goalkeeper, a skilled horse rider and a passionate boxer—essentially, a complete all-rounder. Sainik School surely played a crucial role in shaping Santosh and who he grew up to be, but his father's beliefs, naturally, impacted him as well.

---

1 Col Santosh Yashwant Mahadik was adopted by his maternal grandparents, who lived with their daughter Kalinda and son-in-law. So, instead of his father's surname he always used his maternal grandparents' surname, Mahadik.

In my conversations with him, Tatya bhaiya mentioned a boxing tournament at Sainik School, when Santosh was in his senior house, Rana Pratap. Tatya and his father Madhukar Ghorpade had gone to watch Santosh participate. Boxing was one of the school's most physically taxing games, demanding mental and physical resilience. Santosh had won his match gloriously, but when a friend congratulated Ghorpade, he replied, 'My son's real fight will be against the enemies of the nation at the borders. My son will always be victorious. This is not my son's greatest fight.' Santosh, hearing this, smiled and agreed, saying, '*Zarur hoga* (It will certainly happen).'

I think this incident is illustrative of the kind of conviction and passion that Col Santosh Mahadik was raised with. His father believed his young boy, from the warrior clan of 96 Kuli Maratha,[2] would grow up to be the best soldier the country had ever seen.

Santosh eventually completed his education at Sainik School, Satara, and wrote the entrance examination for admission to the National Defence Academy. To his chagrin, he couldn't get through in his first attempt. So, he did the next best thing and applied to the Yashwantrao Chavan Institute of Science, Satara, from where he cracked the entrance to the Indian Military Academy (IMA) in July 1997. There, Santosh joined the 103-Regular Course. After his passing-out parade in December 1998, he volunteered for the Para Special Forces because he had heard that only the best soldiers made it to these units. Gruelling hours of probation followed, where Santosh was beaten black and blue and forced to starve. There was one week, referred to as 'stress week', followed by escape and evasion exercises. Only after successfully completing all this did Santosh manage to gain entry to the 21 Para (SF) unit.

---

2 The Maratha Kshatriya caste (also referred to as Shahannava Kuli Marathas, 96 Kuli Marathas or 96K), is spread over 96 clans in the Deccan and played a key role in bringing an end to Mughal rule in India.

As an officer of the elite unit, Col Mahadik led numerous successful operations and participated in almost all other ongoing operations across the country for over a decade. But it was in Kargil that his journey to becoming a true Waghnakh began.

As I researched his life and pieced his story together, I realized that Col Santosh Mahadik was always meant to have the left the planet as he did—guns blazing, even as he led his troops from the front.

There was no other way.

~

June 1999
Drass, Point 5605
Kaksar sector
Kargil

Second Lieutenant Santosh Yashwant Mahadik was a newly commissioned officer of the elite 21 Para (SF), or Waghnakhs. The harsh SF training had just been completed and, to Santosh's delight, he had had the opportunity to participate in a full-blown war almost immediately after his commissioning. On 28 May 1999, he boarded an IL-76 Gajraj[3] from an airfield in Assam to Srinagar, along with the 21 Para (SF) assault team, with a full battle load, to participate in the Kargil War.

The battlefield was full of sound and fury. Pakistan's Operation Koh-e-Paima (Mountain Climber), popularly known as the Kargil Operation, had attempted to viciously cut off Indian troops from the rest of India at Leh. Mujahideens and terrorists backed by the Pakistan Army, the Special Services Group (akin to the Indian Para

---

3   The Gajraj is a four-engine multipurpose turbofan strategic airlifter and military-transport aircraft. The aircraft can not only deliver heavy machinery to remote areas but also carries tanks and artillery.

SF) and the Northern Light Infantry infiltrated and occupied the heights that dominated the Zoji La–Drass–Kargil–Leh National Highway, a lifeline for the Indian Army's deployment, sustenance and survival. The idea behind this was to pressure the Government of India to come to the negotiating table on the status of Kashmir. Pakistani generals were hopeful that the operation would cause India to—at the very least—vacate the Siachen Glacier, which India had taken control of in 1984. Ideally, nothing should have gone wrong with such a well-planned strategic operation. But Pakistan had not accounted for the grit, resilience and courage of the ordinary Indian soldier and the overwhelming aggression shown by the Indian Army.

The stage was set for Second Lieutenant Mahadik's debut, against the backdrop of Indian firepower and valour. In the words of his comrade-in-arms, who was with him during the Kargil War, 'From Pandrass to Drass, there was a stretch of twenty kilometres where one could find artillery guns placed so densely that even their tyres would be touching each other. The Indian artillery bombarded the Pakistani side day and night. Explosions grew to a crashing crescendo, with shockwaves causing the ground we stood on to vibrate. The firing was so intense that it resembled the most furious lightning. Santo[4] would find it all fascinating. Often, he simply lay down, watching the sky filled with artillery fire, enjoying the magnificent view. For an outsider, it might have been chaos, but for a soldier, it was all run-of-the-mill.'

As a member of the surveillance-detachment party of the Alpha team of 21 Para (SF), Mahadik was sent to keep track of Pakistani movement. Here, information was key to winning battles. Surveillance teams, such as the one that Mahadik was part of, worked in complete secrecy and silence. He and his team would successfully climb the harsh and steep mountains, often at heights of up to 12,000–14,000 feet, using ropes. These climbing sessions would

---

4   Col Santosh Mahadik was lovingly called 'Santo' by his comrades.

strictly be conducted during the night by making small bases using ropes. Along with his comrades, the fledgling SF officer would sit in hiding at strategic positions for many hours at a stretch during the day. Climbing was extremely tiring, yet the team completed every task given to them. Sometimes, they would hide for days without any food or water, silently observing enemy movement, unbeknownst to anyone. Paras are trained well to hide and attack only when needed. Santosh diligently followed the drill, passing coordinates to his team and directing an accurate artillery strike. The Pakistanis never knew who signed their death warrants.

21 Para (SF) was also tasked with evacuating the ultra-steep Point 5605, named for its altitude in metres, from Pakistani clutches during the Kargil War. Point 5605 dominated the strategically important Neelam Valley in the Drass subsector. At its peak was a group of around thirty Pakistani soldiers.

The entire operation was led by the C Team Commander, Major Deependra Singh Sengar, SM (retd), a daring Para officer known for his excellent execution and well-laid plans. The fire support was to be provided by Maj Ashish Sonal VrC. Mahadik was part of Maj Sonal's squad at the firebase. The battle was fought at a height of around 14,000 feet at a steep gradient. The firebases provided an excellent vantage point from which to view Pakistani posts and had been established in such a manner that they did not fall in the line of direct enemy fire. Continuous Pakistani artillery shelling imposed grave threats but strategically established firebases to provide fire support to the assault parties remained hidden. For Mahadik, it was the experience of a lifetime. Many a time, he was a part of missions where heavy weapons and equipment needed to be lifted to various peaks to establish firebases. It was tough carrying heavy loads. The squad would use ropes to lift them inch by inch at high altitudes in the cold and dark. But the young officer never backed down. The battle ensued for several days but eventually Point 5605 was recaptured and Santosh was part of that glorious victory now registered in the annals of the unit's victories.

Mahadik was also part of another mission to capture the strategic Safed Nallah area ahead of Mushkoh Valley in Drass.[5] This was used by the Pakistanis as a route to supply rations, using mules, to Tiger Hill. Due to the geographical terrain, the area was naturally hidden amidst a series of peaks and was not visible from heights. Using heavy fire, the Alpha team pushed the enemy back and recaptured Safed Nallah. Mahadik was part of every action, from launching mortars at the enemy posts to loading rounds into the 84 mm guns to working as part of the surveillance teams. His men loved him, and his seniors were happy.

By the end of the Kargil War, Second Lieutenant Santosh Mahadik had managed to survive and learn many lessons on that fateful battlefield.

His most important takeaway was that Paras either kill or die—there was no middle path.

~

2001
Operation Rakshak
Kupwara
Kashmir

Brig Saurabh Singh Shekhawat, KC, SC, SM, VSM, renowned as one of the most decorated officers in the Indian Army, brimmed visibly with pride when I told him that I was writing a story about Col Santosh Yashwant Mahadik. I wanted to know about the bond

5 https://economictimes.indiatimes.com/news/defence/20-years-after-kargil-war-remnants-of-pakistan-army-found-in-dras/articleshow/70121063.cms?from=mdr)

they shared as brothers-in-arms. Not many people know that both these daring officers of the SF battalion worked together in many challenging operations.

Brig Shekhawat smiled fondly as he recalled his time with Col Mahadik, SM, SC (posthumous), for whom he had been an immediate senior, team commander and commanding officer at various points in time, and said, 'Santo was a fearless officer. He performed even the smallest task with the utmost sincerity. His operational IQ and tracking abilities were brilliant. He was the finest amongst us. If he sensed a terrorist around, there would indeed be a terrorist around. He loved his spatial-thermal-imaging inspection camera, which he always had around his neck, and he loved walking ahead of all the scouts.'

The Brigadier paused before continuing. 'But you know, he had no airs about him. He was one of the most humble, down-to-earth and polite people you ever met, despite the superior operational skills he naturally possessed. This quality also made him extremely popular amongst his men. They followed Santo blindly. In Para SF battalions, you can't command men if you don't win their hearts.'

My research told me that the two men had worked closely together in numerous dangerous and ultimately successful missions, engaging in either regular unconventional warfare or special reconnaissance operations. Both men from 21 Para (SF) hold a legendary status in the world of SF. To date, many SF operatives seek inspiration from Santo and Shekhu's daredevil operations and fearless attitude.

For the Para SF operatives, life and death depend on chance, a flip of the coin. The men are fully aware of and possess a sense of humour and self-awareness about this fact. In those days, Kupwara was volatile and infiltration activities were at an all-time high. Santosh would often plan ambushes in the nearby forests, hoping to catch terrorists. Every day, Santosh, along with his boys, would apply camouflage paint and hide in the roots of humongous trees or in the branches. Sometimes they also hid under bushes or carpets of thick leaves without moving for hours. In this world, getting a contact[6] is considered a matter of luck. Soldiers might have to sit for weeks, hoping in vain that terrorists cross their path. Still, that is part of the job. One must follow the drill, keep up the patrolling and carry out constant reconnaissance and surveillance operations.

On one such routine day, Santosh received information about a large group of terrorists on his radio link which had secured communication. The input was garbled but the message was that a large group of terrorists was infiltrating the border in small groups. It was at this exact moment that Mahadik noticed them: two terrorists, with their guns slung across their backs and muzzles towards the ground, striding ahead of his ambush position.

They were easy targets—both were clean shots. But Santosh took a chance, letting them walk away in the hope that they would eventually engage with a larger party of terrorists. Tactically, it was a good move. Had he fired, the gunshots would have alerted the rest of the militants. He signalled his men not to take any action and maintain their positions. They waited for quite some time but there was no sign of other terrorists. Santosh began wondering if he had misread the situation and made a mistake in letting the two terrorists go. So, he decided to chase them.

Santosh and his men came out from their hiding positions, moving ahead slowly in pursuit. Suddenly, they heard heavy footsteps behind them. Startled, they hurriedly took cover wherever possible. The SF operatives numbered only six—and they had no idea how many terrorists would be behind them.[7] The situation had suddenly

---

6 When forces engage in firefight with a terrorist
7 Later, it was revealed that about fifteen terrorists were walking behind Santosh and his men. In the 90s and 2000s, infiltration and insurgency were at their highest.

escalated. Santosh and his men had just been sandwiched from both sides.

It wasn't long before firing began. Slogans of '*Nara-e-takbeer*' and '*Allah hu Akbar*' rent the air, along with threats of how the terrorists would roast the kafirs. Keeping his cool, Mahadik contacted a nearby unit and gave them the terrorists' coordinates, requesting mortar fire. The mortars were duly fired, but they began landing amidst Santosh's men. There was instant chaos, which was duly picked up by the Platoon Commander. He radioed Santosh urgently, saying, 'Santo, we hope you're directing the fire on the terrorists and not on yourselves. *'Tumne apne coordinates toh nahi pass kar diye* (You haven't passed your own coordinates by mistake, have you)?'

Coolly, Mahadik responded, 'No, sir, come on! I have passed on the right coordinates but your mortars are crossing me and my boys left and right. You need to work on your shots. Please fire on the terrorists.'

Eventually, two terrorists were killed, and as soon as the mortar firing intensified, the rest of the terrorists fled. One of the soldiers involved in the operation shared his views with me, '*Yeh jihad-wihad na goli chalte hi nikal jata hai inka* (The moment we open fire at the terrorists, all their pretence of fighting for Allah vanishes and they start running for their survival). How long do you think the enemy nation can hire goons and provide funds to fight a war with no purpose in today's times? Though we don't complain, it is not every day that we get firsthand targets to hone our skills.'

This remark was followed by an enormous burst of laughter which brought a smile to my face. It's too bad that legends don't live forever.

December 2001
Operation Santo
Somehwere in Lolab Valley
Kashmir

Many Para Special Forces operatives that I interviewed during the course of writing this book told me fondly about the bewitching beauty of the Lolab Valley. 'Lol-ab' translates to 'beautiful water'. The valley is an idyllic narrow floor on the north-western edge of Kashmir, which evokes contrasting responses from the men on either side, who are sworn to guard it with their lives. One set remains bound by both the written and unwritten rules of combat, those that a typical, professional Army respects. The other set remains bound, if at all, only by the religious and ideological fanaticism that defines their existence. But in the SF circuit, Lolab stands for the 'land of love and beauty', where one finds the sweetest apples, the prettiest girls—and the fiercest militants.

Col Santosh Mahadik led many successful military operations in the Lolab Valley during his years in service, but it was for Operation Santo in 2001 that he was awarded his first gallantry medal as a young captain. The operation was executed with such conviction and precision that not one man from Mahadik's side was injured. It was a huge feat.

The Lolab Valley is strategically and operationally critical for the SF as it serves as the staging area for newly inducted terrorists from Pakistan trying to infiltrate the border in order to enter Kashmir. Here, Mahadik led a nocturnal life, typical of most Para (SF) operatives. Every night he would climb treacherous hills and mountains laden with snow, carrying his Vz. 58, a standard-issue Czechoslovakian-origin gun. His boys from the Alpha team would match him step for step. In standard formations, and during such operations, scouts walk ahead of the rest of the squad, acting as their eyes and ears. As a result of the dangers they brave, they are revered figures with an awe-inspiring reputation in their field.

Just like these scouts, Mahadik always preferred—right from the beginning of his career—to walk ahead of his boys. Watching him was like attending a masterclass: he could show the best scouts how it was done. Ever alert, Mahadik would perform all his duties excellently; he knew the slightest mistake on his watch could cost the lives of innocent compatriots.

Mahadik would effortlessly negotiate the steep incline of the forest floor or climb across perforated crevasses—there was not a twig-break nor a leaf-crunch to be heard. On alternate nights, when the other squad was not on patrol, he led patrols that returned before the first rays of the morning sun had touched the valley floor. He would get up at 2.30 a.m. for his shift and start patrolling. The night would resound with the barking of dogs far away. There were inputs of heavy infiltration in the area. He kept on following the standard drills of laying ambush in target areas, but the lack of success in locating terrorists haunted his slumber. It was the nearby girls' school that troubled Mahadik the most, which was where dogs barked the most.

So, one fine day, Capt Mahadik changed his regular course of action and decided to lay an ambush in the girls' school. He chose a parapet wall as the perfect cover for the ambush in case of an attack. A squad of six SF operatives took cover in strategic positions and waited for the first sign of any movement.

Several hours passed, with the men lying in a prone position, unmoving. Just when Mahadik was on the verge of giving up, he saw them: two men carrying rocket-propelled grenades (RPGs) and striding towards them. The Indian Army, Mahadik knew, did not use RPGs. These men, then, were hostile.

The next few minutes happened in a flash. Capt Mahadik was the first to open fire—and he hit his target perfectly. But even though one terrorist was down, there was still an intense firefight to contend with—more terrorists entered the scene and retaliated with the unmistakable angry flashes from their AK-47s. The firing forced Santosh to drop down, crawl, aim and shoot. But it was clear after a few moments that the group of terrorists was bigger than his entire squad. Pulling out his radio, Santosh relayed a crisp three-word message over the air, indicating to the Company Operating Base that the squad needed backup: *Contact. Standby. Out.*

The terrorists outnumbered Mahadik's team but the well-laid ambush and tight cordon he'd planned provided tactical and strategic superiority to the squad. They could easily fire shots at the terrorists in the open. Col Mahadik killed one terrorist on the spot. Undeterred and disregarding his personal safety he then crawled towards another terrorist who was firing heavily and killed him at close range. Showing exemplary courage, he once again moved towards his squad, in contact with two other terrorists. Under his effective covering fire his buddy Havildar closed in with the terrorists and killed them both. It was a remarkable battle drill. As soon as the enemy realized this, they fled. By the time the backup team arrived in their Casspir,[8] several terrorists had been killed. Four terrorists were killed in that operation, out of which Capt Mahadik had killed two himself

Capt Mahadik was later awarded the Sena Medal for his operation. His citation reads,

The officer's exceptional courage and remarkable battle drill achieved to the elimination of four terrorists and recovery of three AK-47 rifles and one RPG.

~

---

8 A mine-resistant ambush-protected vehicle.

2002
Pogarwadi
Satara
Maharashtra

Life as a SF operative is lived in the present. A commando's life dangles on a thin thread, where a split second can decide his fate forever. A man armed with lethal weaponry, geared to kill and protect, is a fascinating, if stereotypical, image. But that man also has to deal with the toll that an excruciating job can take, as someone who negotiates with life and death on a daily basis.

But there are moments when happiness stops by as well: romance and lovers, marriage, holding newborns for the first time and seeing them grow up.

Capt Mahadik visited his family in Pogarwadi when on leave in November 2002. He was happy and excited when he boarded his flight but never for a second could he have known that his life was about to change forever. In his village, his father was a highly respected man. Many marriage proposals had already come in for Santosh. Until now, he had always managed to dodge the issue, but now that he was coming home, he had no choice but to succumb to his father's demands and at least see the girls shortlisted for him.

But that didn't stop him from protesting. Ghorpade, Santosh's father, had to take his son almost forcibly to a girl's house in the neighbouring village. As they left, a visibly annoyed Santosh cribbed about how he was going to reject the girl in front of her mother and sister. But when he arrived at the house, something almost magical happened. He ended up falling for the girl—Swati—and would later tell her, '*Sadhadi sojhwar swabhav* (Your simplicity won me over).'

Swati was indeed a simple girl. She had just earned her postgraduate social work degree in medicine and psychiatry and was working. During their conversations, both of them realized that not only had they graduated from the same college but they had also walked by each other on campus. In those days, Santosh had been the Student General Secretary of Yashwantrao Chavan Institute of Science, Satara. But then he had joined the IMA and had never seen Swati after that.

Santosh had always liked Swati but the course his life had taken had made it impossible for him to confess his feelings to her. It was destiny, then, that took him back to her, so many years later. As soon as Santosh returned home, he told his mother, 'Aai, I liked the girl very much. But please ask them if she liked me?' His mother could sense his nerves and confusion.

She laughed and said, 'You told us in the morning that you are not interested in getting married, and now you want to rush into it?'

Santosh only smiled and stepped out of the house, perhaps wanting to process everything that had happened to him over the course of a mere few hours that fine morning. He had found 'the one'.

Swati was on cloud nine too. She belonged to a conservative joint Maratha family, which imposed many restrictions on their girls. As soon as their engagement was fixed, Santosh, Swati and her uncle went to Satara to buy the engagement rings. Swati did not like anything in the first shop. After trying out various rings, she wanted to try a different jewellery shop. Her uncle was annoyed by then and snapped, 'What is this, Swati? A ring is just a ring. Just choose one.'

Santosh took Swati aside, saying quietly, 'Don't listen to your uncle, Swati. Keep looking for the ring of your choice. We will go to every shop here until you find exactly what you're looking for. Don't stop looking.'

Swati was captivated. All her life, she had always been told to seek permission from her elders for every little thing. Now, she knew her life was finally going to change. Her fiancé was a supportive and liberal man.

Santosh and Swati were engaged on 14 December 2002 in a simple Maharashtrian ceremony. To this day, Swati remembers precious moments from their courtship: 'I didn't know anything about the Army, forget the Special Forces. But the only thing I saw was how shyly he smiled every time he saw me. I noticed that he couldn't take his eyes off me. But whenever I looked back at him, he would quickly look away. Right after our engagement, he fell from his terrace and injured himself. He had to be admitted to a hospital in Pune for some time. I was working in Pune then too, so I bunked office every day to visit him at the hospital. His face would brighten, and he wouldn't be able to stop smiling. We would talk for hours. I never realized he was a fierce commando. To me, he was always a man hopelessly in love.'

I watched Swati's eyes light up and heard the happiness in her voice as she spoke of Santosh, of the memories of her life with him. But it had been difficult to get her to talk, to release some of those memories and share them with me. Listening to her story, I could understand why. There was a lot of pain in remembering them.

Soon after his recovery and their engagement, Santosh returned to his unit, which had been posted back to the Northeast from Kashmir. Swati and Santosh would write letters to each other to try and bridge the distance between them. Those letters would be a symbol of their love for each other.

~

2003
Assam

Lt Bhupendra Manohar Dhamankar was received by a fit, well-built officer with a ramrod-straight posture, whom he later came to know as Capt Santosh Mahadik. Dhamankar was awestruck by his quiet, polite and humble demeanour. It was nothing close to what he had thought Mahadik would be like. He knew of Mahadik's achievements so far: he had conducted some excellent operations and was a recipient of the Sena Medal.

Dhamankar had been sent to the elite 21 Para (SF) unit for probation and was about to start the same journey under Capt Santosh, then a troop commander in the Alpha team.

Capt Dhamankar (retd) told me, 'It was under Santo sir that I learnt the most. He was so humble and down to earth, it was hard to believe that I was talking to a decorated officer. He always said that as an officer, one must always do more work than the boys one commands. An officer must prove himself to his boys and prove his worth through his actions. He must perform his duties with the utmost devotion and honesty. Only then will the men under his command accept him on their own and blindly trust him and put their lives in his hands. They are not going to follow the officer until they truly believe in him.'

Capt Dhamankar told me the story of Mahadik taking clean shots at terrorists who were leaping over nallahs. They never knew where 'the killing shot' came from, he said, adding, 'It's a rare skill. It's difficult to focus and shoot a moving target, yet Santosh sir did it easily.'

While Santosh was much loved by his men on the professional front, he also took a step in his personal life and committed to Swati. They were married on 4 July 2003. Swati recalled how Santosh carried her in his arms and put her in their car during the vidaai:[9] 'I was the eldest one in the family and the first to get married. Everyone was sobbing, and I was emotional too. I was becoming part of another family, leaving my own. My heart was in turmoil. There was a whole queue of uncles, aunts, cousins, relatives and friends all sobbing, all of them waiting to meet me. There was a lot of heavy emotion in the

---

9   The Hindu ceremony by the bride's family of bidding the bride farewell after the wedding.

air. Suddenly, Santosh picked me up and put me in the car. He said, "That's enough now. I'm taking my bride home." That broke the air of sadness and, just like that, everyone began laughing. You see, he didn't like to see me cry. *Vidaai bhi theek se nahi karne di* (He didn't even let me complete my vidaai properly)!'

Once they were married, Santosh brought Swati to Assam, where they set up their two-room accommodation. It was a home that opened its doors to everyone from the beginning. Swati is an excellent cook; her speciality was Maharashtrian delicacies like puran poli, faral, chakli, chuda, karanji and sev bhakadi. You just had to name the dish and she would be able to whip it up. Whenever the couple decided to throw a party, or if bachelors banged on the door of their little fauji accommodation in the middle of the night for something to eat, Swati would merely smile and, within minutes, would be in the kitchen cooking up a storm.

Capt Dhamankar told me, 'I am a Maharashtrian, and whenever I would get a craving for Maharashtrian food, I would beg ma'am to cook my favourite dishes. But it was just not me she was kind to; she was warm and welcoming to all the bachelors who would invariably be found at sir and ma'am's home with their demands for home-cooked food. She fulfilled all their food requests with a smile.'

Swati Mahadik is now a serving officer in the Indian Army herself. When asked about her memories of those early days of marriage, she chirped, 'Those were beautiful days. We were newly married. I was young and naive. In the excitement of joining him, I left my job and joined him in the Northeast, but he hardly ever stayed home. He would vanish for months, only to come back home for a week. I would be scared and worried because when he was on his missions, he would not even contact me nor mention anything. But yes, whenever he was home, I would feel alive.'

The newlyweds were very much in love. When Swati got angry after a fight, Santosh would sing '*Dilbar mere*' to her till she gave in to his relentless charm. Santosh also loved dancing with Swati. They loved grooving to '*Dhadkan zara ruk gayee hai*'. It wasn't long before Swati was pregnant.

Kartikee, their baby girl, was born on 27 December 2004 in a military hospital. Santosh was overjoyed at her arrival, unable to stop dancing. He finally had a family of his own; he was a part of something so magnificent that he could hardly believe it.

~

2004–12
Assam, Manipur, Delhi, Bengaluru
Operation Hifazat and Operation Rhino

Col Santosh Yashwant Mahadik was a man of few words. He ardently followed the Chetwode Motto: *The safety, honour and welfare of your country come first, always and every time. The honour, welfare and comfort of the men you command come next. Your ease, comfort and safety come last, always and every time.*

Mahadik loved his nation and his men. They always came first, even before his little family, which now comprised four people, following the birth of his son, Swaraj (named after Shivaji's dream), in 2010.

He played an active and important part in almost all the ongoing operations at the various critical border areas of India. It is not possible to write about all the contributions Col Mahadik made throughout his service, but whether we are talking about Operation Vijay in Kargil, Operation Orchid in Arunachal Pradesh, Operation Rhino in Assam, Operation Rakshak in Jammu and Kashmir or Operation Hifazat in Manipur, he has been through all of them.

Col Mahadik was also an ace combat diver.[10] In 2001, he completed his Army Combat Underwater Diving Course in Cochin. The military's combat-diving course is one of the most challenging; many say it is even tougher than SF probation. There is no margin for error underwater; there are no anchors for support—only the dark mud, the cold, and unknown threats. Operating underwater is simply a test of human resilience and instincts. Mahadik swam and trained for hours in the sea. From learning how to carry out underwater rescue operations to demolishing enemy bridges and ships, he excelled in the course and earned the coveted diving badge of a Combat Deep-Sea Diver.

He volunteered for challenging underwater missions to rescue people or recover dead bodies from rivers or seas. Such missions are categorized under 'high-risk operations', which may take several days to complete. The combat diver faces a plethora of challenges, usually beyond ordinary human capabilities. For example, he may have to manoeuvre through deep dark waters and find a way through after checking everything with his own hands. He also risks facing dangerous creatures like alligators or sharks and losing his way altogether.

Col Mahadik saved numerous lives and recovered multiple bodies and weapons from many 'missions impossible'. A diver told me about a risky and emotional mission in which Mahadik had volunteered to recover the body of a fellow soldier who had drowned in the

---

10 Combat diving requires a higher level of skill, courage and daredevilry than regular deep-sea diving. Combat divers are among the most coveted SF soldiers in the Indian Army. Every mission or exercise they undertake poses a real, life-threatening risk. Watch the American action-adventure drama film *The Guardian* (2006), directed by Andrew Davis, starring Kevin Costner and Ashton Kutcher, to get a small taste of the risks involved in being a combat diver.

Brahmaputra River, considered one of the mightiest and fiercest rivers of India, during a combat exercise in Assam. Joint teams comprising the National Disaster Response Force, local police, Indian Army and others were deployed. After a few days of searching, all of them unequivocally opined that it was going to be an impossible task to carry out any further search operations for the soldier's body against the excruciatingly piercing currents of the dark river.

The SF were called upon to help and Col Mahadik (then a major) volunteered for the operation. He ended up recovering the body of the slain soldier who had sacrificed his life in the line of duty and was simply awaiting an honourable funeral in the presence of his friends, family and the nation. The body was later given to the family members who could grieve the death of their loved one properly—all because of Col Santosh Yashwant Mahadik, SC (posthumous), SM.

Col Mahadik was also a skilled marksman who led many thrilling operations in the Northeast. As a team commander in a village in Manipur, he killed a terrorist belonging to the People's Liberation Army of Manipur (PLA) during one such operation. He laid an ambush for two continuous days on a mountain, hoping to bait terrorists. The ambush was based entirely on intelligence inputs, which play a pivotal part in any battle. On the third day, when twelve men under his command were thinking of returning to the base, they saw terrorists moving at another point in the nearby area. In the midst of a running firefight, in which the Indian troops ran down an alley behind the fleeing terrorists, Maj Santosh Mahadik decisively shot down a terrorist—a testament to his extraordinary marksmanship.

There are numerous other missions that shed light on this man, a hero of few words, who gave his all to his country. The collective contributions of the SF, the rest of the Indian Army and other forces resulted in the Northeast becoming relatively free of the terrorism, kidnappings and violence that once characterized it.

There are many similar stories of other brave men who worked silently to keep us safe. We will never truly know what it takes to keep the freedom and sovereignty of our great nation alive.

When I looked for more people to speak to while researching Col Santosh Mahadik, entire *teams* came forward to talk about his legend and how he touched their lives. I can only imagine how heartbreaking and unbelievable it has been for his loved ones to live with the fact that he is not among them.

~

2013

Kupwara

Kashmir

In July 2013, Col Mahadik was posted to the 41 Rashtriya Rifles[11] as the 2IC[12] of the unit; he took over full command in July 2014. The Rashtriya Rifles (RR) are highly professional military battalions which see a high turnover of manpower throughout the year. Soldiers from different arms and units are attached to the RR, contributing their expertise to create a military outfit that is on a par with other world-class forces. But even under these circumstances, it is rare that an RR battalion is commanded by an SF officer. SF units are reluctant to share their homegrown trained officers, who have killer instincts and the skill to command the entire unit, with another unit when they face a scarcity of men already. As a result, it was a thrilling moment when 41 RR received Mahadik as its commanding officer in 2014.

---

11 https://twitter.com/spokespersonmod/status/666998762645000192?l ang=gu

12 2IC: Second-in-Command in the chain of hierarchy in a unit, immediately after the Commanding Officer.

Col Mahadik lived up to the expectations of his seniors. Since he had been working as 2IC for quite some time with the unit, he was well acquainted with the operations. As commanding officer, this kind, soft-spoken, compassionate, and empathetic officer ensured that he gave his best at every level. He would maintain a diary where he wrote down the problems of his men and tried to solve them on priority. He knew all his men by name, and his innate kindness set him apart, because a Rashtriya Rifles commanding officer deals with an unimaginable level of workload and pressure.

The locals, too, waxed eloquent on Col Mahadik's generosity and the compassion he had for Kashmiris. He was posted in the Kupwara area. He had come full circle: it was, after all, the place where he had first begun his career as a young Para SF operative, of which he had many fond memories.

Shamim Khan, who owns a restaurant in Kupwara and who catered many official banquets for Col Mahadik, told me fondly, 'Besides being a brave soldier, he was also an intellectual who wished to capitalize on the natural beauty of Kupwara and revive tourism in the area. He reconstructed old monuments and organized football and cricket matches to build bridges with the local community. He felt the revival of tourism would not only provide employment to the youth but would also attack the roots of radicalization. He sent local youth on tours to Jaipur and Rishikesh to learn about village tourism and white-water rafting. He would personally counsel ex-militants and guide them on the path to a new life.'

Khan paused before continuing, 'Now, when we go to meet high-ranking generals of the Indian Army regarding the various problems we face, we always talk about how Santosh Saheb would have definitely returned to the area as a senior officer had he survived his fate. It was a loss to Kupwara.'

The impact of his Sadbhhavana initiatives have been so deep that even the press release by the Government of India after he was killed in action especially mentions this:[13]

Colonel Santosh Yashwant Mahadik, SM, a dynamic officer from the Parachute Regiment (Special Forces) was commanding 41 Rashtriya Rifles from July 2014. Col Santosh with his inspirational leadership and humane skills and altered the narrative of kupwara town through successful OP SADBHAVANA initiatives in education, sports and healthcare.

When I talked to Tatya Mahadik, Col Mahadik's elder brother, he said, 'Just a few months ago, before he made the supreme sacrifice, my family and I were touring Jammu and Kashmir. When Santosh heard, he insisted that we visit his unit as well. We were escorted by a squad of soldiers that suddenly appeared out of thin air, almost immediately after I put down the phone. The following day, he arranged for us to explore the natural beauty of Kupwara, which, I would say, I found far superior to Sonmarg. Our guide was a surrendered terrorist. Honestly, my wife was worried, but when I asked Santosh about trusting a terrorist with his family, he smiled and said, "Yes, absolutely. If we talk about the need for change, then militants need to be attached to the mainstream population and given ways to live their lives."'

Tatya also mentioned the huge hoarding that Mahadik had had made and placed at the entrance—it had the words *Every Soldier Is an Officer, Every Officer Is a Soldier* printed on it. It clearly reflected his vision and thoughts about his unit and his men.

Wars are brutal and unfair. They take away our best, our dearest, our liveliest and brightest, leaving a deep void within us. Still, on

---

13 https://pib.gov.in/newsite/PrintRelease.aspx?relid=135802

the brighter side, wars and the legends that emerge from them allow us to witness the pinnacle of human valour and spirit. For Mahadik, there was more to come.

~

9 November 2015
Manigah
Kupwara
Kashmir

Col Santosh Mahadik, Commanding Officer, 41 RR, was having lunch with Col Amitabh Walawalkar (retd), the commanding officer of a neighbouring Rashtriya Rifles unit. It is rare when two SF Officers command adjacent RR units *and* happen to be course mates. Perhaps God knew the end result and had planned everything beforehand. But Mahadik and Walawalkar's association went back a long way. They had taken the Antim Pag (Last Step) in the IMA together, worn the coveted Balidan[14] badge and had received the maroon beret together to become Wagnakhs as well. They had been to each other's homes, and their mothers adored them both equally. Back then, nobody had any idea that one friend would, one day, bear the unimaginable grief and sorrow of carrying his beloved friend's dead body to the family that was almost like his own.

Col Mahadik briefed Col Walawalkar about the ongoing cordon-and-search operation based on specific information about the presence of terrorists in the nearby Manigah forest of Kupwara

---

14 A badge of sacrifice awarded only to the SF personnel of the Parachute regiment after they have spent a year with a Para (SF) battalion, or six months if the battalion is involved in active duty. Col Megh Singh of the famous Meghdoot Force first proposed the adoption of the Balidan insignia.

district. Various other parties from different units were involved in the operation to remove all chances of the Lashkar-e-Taiba terrorists escaping. Col Mahadik would be leading the entire operation. The word 'leading' was used in a literal sense of the word, for, here too, Mahadik wanted to physically lead the search party himself. As commanding officer, he was also responsible for the coordination and planning among all the teams involved. If the mission was led by him, he knew there would be a greater chance of success in a sensitive operation that was also vague in the absence of any concrete information.

While the search operation was under way on 13 November 2015, reports of a quick response team (QRT) led by a lieutenant colonel coming under heavy fire in the Manigah forest were relayed to headquarters. Two soldiers were injured, although not fatally. The attack confirmed the presence of foreign terrorists in the nearby forest, with a strong chance of future havoc.

It was always a cat-and-mouse game between the terrorists and the Army in Kashmir. While there is never any doubt that the Army will eventually find and neutralize the threat, it is also clear that the infiltrators will cause as much harm as humanly possible before they can be eliminated.

On 16 November 2015, reports came in about a search party from an infantry unit exchanging fire with the terrorists in the village of Subaya near Manigah forest. There was information that a Major had been injured along with a jawan. Since security cordons within the forest areas had been tightened, it was obviously getting difficult for the terrorists to manage their supply chains. They were thus coming into civilian areas and mingling dangerously with local populations. It was a sign of their desperation, along with the indiscriminate firing from their hidden positions, even as troops on search operations were now out in the open looking for them.

Col Amitabh Walawalkar (retd) remembers how desperately Santo wanted to eliminate the terrorists, and how he had told Santo at their last lunch meeting, 'Tell me if my unit can do more. The information is not secure, and the area is enormous.'

Col Mahadik had replied with his signature calm, 'Yes, I will let you know. Several parties are already on the prowl but I think now is the time to launch SADO [search-and-destroy operation]. We need to accelerate the rate of the operation and eliminate the terrorists. Otherwise, our men out in the open will be at risk.'

Col Walawalkar had smiled and said, 'Okay! But Santo, you must take it slow. You are a commanding officer—your planning is more important than prowling.'

Col Mahadik had smiled back and responded, 'I plan to leave the Army after my command and open a coaching institute in my area for youngsters who want to join the Army. I want to contribute to my area and help my people. I have served my nation, and now I must serve the people of my village. But as long as I am here, it is my duty to lead my troops from the front. Either way, you know I am the best tracker here. I know how to pin down the bastards.'

The two friends had hugged each other and gone their way. Neither had any idea that this would be their last meeting.

~

17 November 2015
Operation Santosh Manigah
Manigah
Kupwara
Kashmir

Mahadik was relentless. He knew the dangers that lay ahead of him: his men were injured, and the terrorists were roaming freely, taking advantage of the difficult terrain. He launched SADO. The operation is now registered as Operation Santosh Manigah, a glorious page in the history of the 41 Rashtriya Rifles.

He met his local sources and extracted information about the likely location of the terrorists in the forest. Along with his QRT, Mahadik moved to the suspected site, dangerously close to the terrorists, and started deploying his men. Deployment is usually the riskiest aspect of any ambush because one is out in the open and can easily become a target. That is precisely what happened.

Mahadik and his team were spotted almost immediately by the terrorists, who were hidden at an elevated position and covered in thick foliage. The first round of bullets was fired—that first burst hit Col Santosh Mahadik in the front and he sustained multiple gunshot wounds.

Grievously injured, but still striving to live by his principles to the end, Mahadik's first instincts were to save his men. He immediately manoeuvred to outflank the terrorists, engaging them with retaliating fire. In doing so, he deliberately provided a window to his troops out in the open to take cover. Mahadik, the commanding officer covering for his men, did not demand an evacuation and continued fighting until his men took cover. Then, someone from his party informed the headquarters via their radio set: '*CO saab ko goli lag gayi hai* (CO sir has been struck by a bullet).'

That was the end.

Chaos ensued. The commanding officer is a revered figure in the unit, a father to all soldiers under his command. The news of Mahadik falling had a devastating effect, galvanizing all parties beyond the call of duty to evacuate him even as bullets flew from all sides.

Col Santosh Mahadik was evacuated from the forest to the village by his buddy, who found Col Walawalkar awaiting his brother-in-arms. The news had reached him and he had rushed to the area. Mahadik was driven to a nearby military hospital but was soon airlifted to the base hospital in Srinagar, 85 kilometres away.

Santosh was lying peacefully, his head in Walawalkar's lap. He looked as though he were sleeping. He had succumbed to his injuries by 5 p.m. It had been a perfect plan, but it was an even more glorious death—a supreme sacrifice made keeping in mind the highest principles and traditions of the Indian Army.

Swati was in Udhampur, around 265 kilometres from Srinagar. She had already seen the news flashing on the television. Frantically, she had tried several times to contact his unit until she was finally able to reach the adjutant.[15] She asked desperately about the nature of his injuries. But when she was told that Santosh had been shot seven times, there was nothing to say. She hung up the phone quietly.

~

19 November 2015
Pogarwadi
Satara
Maharashtra

Lakhs of mourners gathered in Pogarwadi, Satara, to pay their homage to the brave son of the soil. Even the then defence minister, Sri Manohar Parrikar, and the chief minister of Maharashtra at the time, Sri Devendra Fadnavis, were present. Mahadik's band of brothers in their uniform of olive green and maroon berets accompanied the casket, wrapped in the national flag.

Col Santosh Yashwant Mahadik was given a twenty-one-gun salute and cremated with full military honours.

He was awarded Shaurya Chakra posthumously. His citation reads,

> *Colonel Santosh Y. Mahadik made the supreme sacrifice keeping with the highest tradition of Indian Army while ensuring safety of the troops he commanded and personally led from the forefront.*

---

15 A regimental staff officer whose duties are to assist the CO in the training, administration and maintenance of discipline in the unit.

He was survived by his wife, Swati Mahadik, and children, Kartikee and Swaraj. Swaraj was too young to understand that he had lost his father.

Days later, when Swati received her husband's phone along with his other belongings, she read the last message on it. It was from one of his friends who had asked Santosh to join him for a trip abroad during the upcoming leave period. Mahadik had replied:

7 December to 7 January is dedicated to my lovable, devoted and innocent wife. So apologies, buddy, but I cannot come.

Capt Swati Mahadik strove hard and went through rigorous training to wear the uniform which mattered the most to the love of her life. She refused any leniency offered to her and trained like all other cadets. Kartikee and Swaraj had to be away from her during this period. But they now live together as a family. Kartikee has done exceptionally well. She has completed her arengetram (graduation) in Bharatnatyam and is a professional dancer, even performing internationally. She is also the table tennis captain of her school and has won many senior championships. By the time this book is published, she would have already sat for her twelfth standard board exams. Swaraj is in class seventh. He loves horse riding and is learning classical vocal music. Capt Swati Mahadik is giving her best to their children.

Her husband and her children's father is missed terribly by the small family he has left behind.

～

It has been seven years since Col Santosh Yashwant Mahadik left for his heavenly abode, but his glory still grips the mind of the people of this country. His operations and never-say-die spirit have inspired many SF operatives.

This chapter is based on interviews with Col Mahadik's wife, Capt Swati Mahadik; his brother, Tatya Mahadik; his course mates; Shamim Khan, his close Kashmiri friend; his brothers-in-arms from 21 Para (SF) Colonel Amitabh Walawalkar (retd) and Maj Bhupendra Manohar Dhamankar retd; and many other comrades who are still serving and want to remain anonymous.

His brothers from 21 Para (SF) miss him dearly. They have named their diving tank 'Santosh'. He is still a part of their mess conversations. Had he been alive he would have been gearing up to take over a brigade by now or chosen an alternate path—either way, a great loss to the unit. His legacy is passed on to all new Wagnakhs joining the unit. The unit keeps in touch with the family and takes pride in all the achievements of Capt Mahadik and her children. Needless to say, the band of brothers from 21 still stand like invisible guardians behind them. They went the extra mile to enable me to write this story in his memory. I hope it immortalizes his tales of bravery for many generations to come. Jai Hind!

2

Brigadier Saurabh Singh Shekhawat, KC, SC, SM, VSM: The Man, the Myth, the Legend

A LONE MI-17V5 helicopter was in the process of take-off, its rotor circling frantically and making a deafening noise. Suddenly, above the din, a cry arose, *'Bajrang Bali ki jai* (Hail Lord Bajrang Bali)!'

It was the war cry of the combat free fallers[1] who were about to board the aircraft. As the helicopter lifted off the tarmac, smoke trailing from the exhaust. Inside the aircraft, SF soldiers sat heavily laden with their ram-air parachutes[2] on their shoulders, combat rucksacks on their laps and—in some cases—on their chests.

---

1   High-altitude military parachute specialists who skydive with personal- and squad-support weapon equipment and infiltrate the denied enemy area to carry out a specialized covert operation
2   Military free-fall advanced parachutes with a multi-mission, high-altitude parachute delivery system that allows personnel to exit at altitudes between 3,500 and 35,000 feet.

Combat-ready armaments—from 7.62 AKM rifles to bulky 84 mm rocket launchers and mean 7.62 mm general-purpose machine guns called PKMGs—were on full display. Navigational equipment like altimasters, magnetic compasses and GPS were checked one last time. One could feel the adrenaline in the air.

For readers not in the know, this wide array of weaponry and equipment is carried by the Indian Army's SF operatives while performing combat free falls—the specialized insertion technique of jumping from an aircraft from a height of 25,000 feet or more and stealthily gliding to the target using military parachutes with a full combat load. Combat free fallers carry out covert military operations in cases where the target cannot be destroyed, using naval or land forces or even an aircraft.

Today, there was something unusual, even unfamiliar, lingering in the air, mingled with the heady rush of anticipation.

Near the pilots' cabin, one trooper stood wearing only a harness over his uniform. He was flanked by another paratrooper wearing a helmet, advanced diving goggles and a bulky parachute with many rings on it. The man in the harness was checked and, in no time, the second paratrooper hooked the first man, who had no parachute, onto his chest. They were ready for tandem skydiving.[3]

This menacing-looking second paratrooper was more than just a combat free faller. He was a highly accomplished tandem master,[4] a

---

3 A form of skydiving where a student skydiver is connected to an instructor via a harness. The instructor guides the student through the whole jump from exit through free fall, piloting the canopy as well as landing.

4 A combat free faller who can carry a passenger (not a free faller) hooked up to him in tandem mode and wearing a tandem parachute.

tandem master instructor[5] and an accelerated-free-fall instructor[6]—
all hard-earned titles in the coveted world of paratroopers. One
needs hours of practice to be a paratrooper, hundreds of jumps to
qualify as a combat free faller and several hundred jumps to train as
a tandem master. But when it comes to earning the title of tandem
master instructor and accelerated free fall instructor, one might as
well forget sleep during the training period. Each stage becomes
progressively more difficult to master, with growing responsibilities.
That's how high the stakes are in the clandestine world of military
free fallers.

The man in the harness was Col Saurabh Singh Shekhawat[7].

Col Shekhawat stood upright like a bolt, with the passenger
hooked onto his chest and heavy parachutes strapped onto his
shoulders. To those watching, the men resembled that original fearless
free faller Lord Hanuman, who flew effortlessly, not with a parachute
but with an entire mountain on his back. This is what makes him
perhaps the most favourite god of Indian combat free fallers, who
provides them with hope during tough times.

---

5   An instructor authorized to train tandem masters.
6   The most advanced skydiving instructor rating. The instructor has
    solid control over his own free fall skills and awareness. He is tested on
    both flying skills and teaching abilities. AFF Instructors assist students
    into learning new manoeuvres, i.e., pull sequence, stable freefall, turns,
    forward motion, track, instability drills, etc.
7   The particular event occurs at a time when he was an officer at the rank
    of a Colonel. He was promoted and is a Brigadier rank officer now.

The lights turned green. There was a loud beep. The master dispatcher[8] confirmed from the spotter[9] that he had spotted the drop zone.[10]

Within seconds, the paratroopers—each heavily burdened—leaped out of the aircraft, speeding towards the earth. Behind them came Col Saurabh Singh Shekhawat, carrying another trooper on his chest. The scene seemed right out of a high-voltage Hollywood adventure movie.

Shekhawat paused at the door, turned inwards, arched his body and, whoosh, out the pair went into thin air! Then, precisely after five seconds, he removed and threw up the drogue chute[11] of his tandem parachute. Soon, the pair was just a tiny dot in the sky below the helicopter. Once in the air, Shekhawat did his handle check drill, signalled the passenger hanging below to ease out and enjoy the fall at 230 kmph; at 5,000 feet, he released the drogue. In a flash, the main canopy came out and the pair was suspended under a huge 370 square feet canopy.

---

8 Responsible for ensuring that the correct standard operating procedures are employed in a parachute drop from the emplaning stage until the trooper exists the aircraft.

9 Responsible for the visual identification of the drop zone and ensuring that it's safe to commence the parachute drop. The troopers jump on his command.

10 A place where parachutists or parachuted supplies land. It can be an area targeted for landing by paratroopers, or a base from which recreational parachutists and skydivers take off in aircraft and land under parachutes.

11 A drogue parachute is a parachute designed for deployment from a rapidly moving object. It can be used for various purposes, such as to decrease speed, to provide control and stability or as a pilot parachute to deploy a larger parachute. Vehicles that have used drogue parachutes include multi-stage parachutes, aircraft and spacecraft recovery systems.

Every soldier, technician, doctor and trooper stood looking up at the sky with bated breaths—they were witnessing an unprecedented event. It was the first time in the history of the Para Training School (PTS) that a tandem skydiving jump[12] with full combat gear was being performed. That was also the first jump with a passenger in the history of the Para SF in India. The two free fallers were not just performing any ordinary free-falling exercise—they were creating history.

This dive marked another feather in the cap for the Indian Para SF and also in the history of the PTS which would now have its own Tandem Instructors. These commandos can safely transport into the battlefield any specialist—from doctors and demolition, bomb-disposal or communication experts to sniffer dogs and lethal snipers—critically required for the success of an operation. This ability to enter into heavily guarded enemy territory, with their valuable cargo, undetected by advanced radars, provides a flexibility unmatched by any other infiltration mechanism.

When the people on the ground spotted the pair hanging below one giant parachute, a cheer went up. However, it was soon observed that the parachute was travelling faster than the ram-air canopies they had jumped from. And since the speed was too fast for a safe landing, it looked like it would result in a severe crash landing.

But what the people watching did not know was that this time around, the tandem master in question was Col Saurabh Singh Shekhawat—a three-time Everest climber and a legend in the world

---

12 A tandem skydive is a skydiving experience where two people jump out of an airplane together, strapped to one another during the entire descent. The connecting straps make it so that one person is floating above the other during the free fall, with one's back against the other's front. Only a certified tandem instructor can carry out tandem jumps.

of the Special Forces, one of the most decorated serving officers in Indian Army. Courage and resilience were his inherent characteristics.

Everyone else may have lost hope, but the Colonel had not. He ordered his buddy, 'Legs fold!' At 1.5 feet above the ground, the Colonel pulled his toggles down with a loud grunt. The parachute was controlled, enabling the pair to softly land on the ground. History had been made.

Applause and cheers erupted among the rest of the boys. The Subedar Major ran to congratulate Col Saurabh Singh Shekhawat, who stood by nonchalantly, as if nothing had happened.

~

Brigadier Saurabh Singh Shekhawat, KC, SC, SM, VSM

When I asked him about all the attention he received, Brig Shekhawat said shyly, 'Ma'am, frankly, it does not matter. I don't even want it. All my life, I have only given one proper interview to Times Now. It was a one-hour interview for a book, and apart from that, I've made hardly two to three appearances in public for a short duration. They were all done on Army orders, but every other day, I would find a YouTube video with some or the other absurd fact about me and repetitive images. They don't even take permission nor check the facts. I am just a normal soldier who serves his nation and his people, like thousands of others. My duty and loyalty to the maroon beret pushes me to challenge the impossible. The rest is irrelevant.'

Brig S.S. Shekhawat is someone who can comfortably be called the poster boy of the Indian Special Forces, a man with many crowning accomplishments. Famous for his sense of adventure and grit, he has been known to conquer high mountains and eliminate terrorists under the deadliest circumstances with equal ease. Perhaps this explains some of the mythology that has grown around this figure.

It took me several months to persuade him to tell me about his experiences and then to research, interview and write the story. I felt it was crucial to procure the complete picture for the benefit of the youngsters of our nation and SF operatives who idolize him and aim to achieve more in their lives. Perhaps the prayers of millions of his fans who idolize him have pushed the gods to bestow me with the rare honour of writing his complicated story. He is not just of an SF operative but of an epic mountaineer, ace paratrooper, the one whose every frame of life is inspiring.

How does a storyteller even cover so many aspects in just one book?

Athletic and handsome, with a chiselled frame and muscular body, even in his fifties he can shame any movie star with his looks alone. Brig S.S. Shekhawat's olive-green uniform is studded with peacetime gallantry awards and medals: a Kirti Chakra for Operation Loktak Vijay on 26 January 2009, a Shaurya Chakra for summiting Mount Everest on 15 August 2001, a Sena Medal in Operation Hifazat on 26 January 2008, a Vishisht Seva Medal for leading the first-ever women's team of the Indian Army to Mt Everest on 26 January 2006, a Balidan badge, a Commando Instructor badge and a paratrooper wing. Did you know that he is one of the highest decorated serving officers of the Indian Army?

Brig Shekhawat's average day begins with a 4 a.m. session of physical training with his troops. After this, he goes horseback riding—a sport he picked up when he was just four years old. He concludes his exercise with a swim. By the time the clock strikes seven, he is done with his morning routine and gearing up for office. It is a strictly regimented routine but, then again, discipline has always been an important part of his life.

~

1982
Brigadier Shekhawat's village
Alwar
Rajasthan

Saurabh Singh Shekhawat was eleven years old. He sweated profusely as he loaded the harvest onto the tractor. Only that morning he had finished cutting the crops along with his cousins—now, they had to load hundreds of kilograms of produce before Dadu Sa—Maj Pratap Singh, his paternal grandfather—came along. He was a revered figure in their family.

The Shekhawats belong to the prominent Rajput clan of the Kushwahas, tracing their ancestry to Kush. The elder son of Bhagwan Sri Ram, Brig SS Shekhawat's ancestors are natives of a small village in Alwar, Rajasthan. Several generations of Shekhawat's family have served in the Army. His grandfather, Maj Pratap Singh, fought in the Second World War and his great-grandfather, Capt Shivnath Singh, fought in the First World War. The women in the family looked after the home front, with their men off participating in wars in faraway lands. Often, they had to fight to protect themselves, their children and the family land. Courage and audacity ran in their blood.

Maj Pratap Singh served in the Alwar State Force and then in 2/9 and the 3/11 Gorkha Rifles of the Indian Army after Independence. He was a strict disciplinarian and wanted all his children and grandchildren to be tough and hard-working. Every time any of his grandchildren would be home for their vacations, he ensured they never had a free second. Both girls and boys were given arduous tasks to complete and were frequently exposed to the harsh realities of life.

Saurabh was the eldest grandson, so he was always given extra work and punished harder too. But he never felt bad about it and still remembers those times with a lot of pride. He says, 'My grandfather has played a huge role in shaping me into who I am today. He made me strong and tough. Those were my formative years, and Dadu Sa stood tall in my life with his influence on me. I remember him taking out his binoculars, standing on our roof and surveying the area. When he found cattle or camels grazing in our fields, he would make us run behind them—we would run two to three kilometres daily this way. Sometimes we would fight with the cattle owners too. My average day was spent milking cows, bathing cattle and collecting their dung. I would also help in cutting the crops, stuffing the mounds into hundreds of sacks and then carrying those sacks one by one along with my cousins to load onto the tractors. Youngsters today might find a story of growing up amidst cattle hard to believe. I was fortunate that my life on the farm and the time spent with my grandfather during my formative years instilled in me the belief that nothing was impossible.'

Brig Shekhawat's childhood friend, Ashwini Singh, an IIT Delhi alumnus who currently works as a software engineering executive with Bank of America in the United States, shares, 'Saurabh always stood out. We met in class eight, and I remember him being the tallest guy in the class. We were popularly referred to as the Jai–Veeru[13] jodi [duo] in school—I was the brains and he was the brawn. He was rugged and rough … I remember running behind camels—all of us would give up in some time, but Saurabh would eventually catch a camel and ride it. We were regular teenage boys—we would be pulled up for our naughtiness, but that never deterred us from planning new adventures.'

Brig S.S. Shekhawat's mother, Dr Shraddha Chauhan, is the epitome of elegance and poise. She says with great pride, 'We led a tough life in those days. Both the men as well as women in our family are accustomed to hard work. We are restless souls, always in pursuit

---

13 An iconic best friend duo from the movie *Sholay*.

of new horizons, be it in the field of academics or on the battlefield. My family does not know how to settle for less. There were times when I crossed rivers with all three of my children so that they could attend school. There were times when Saurabh slept in the fields with snakes crawling around, and rather than running away, he would go catch them—and I would encourage it. Saurabh has grown up in a family where fearlessness, courage and the quest for adventure are not considered special traits; they are just typical Shekhawat traits.'

Saurabh Singh Shekhawat was always meant for the Special Forces—wearing the maroon beret, living among high peaks and flying free in the open skies. It was destiny.

Nobody was surprised when he cleared the Combined Defence Services[14] exams in 1992. The family was on cloud nine. Saurabh had carried forward the family legacy.

~

11 June 1994
Indian Military Academy
Dehradun
Uttarakhand

The historic Chetwode Drill Square at the Indian Military Academy (IMA), Dehradun, was abuzz with excitement during the passing-out parade. Indian and foreign cadets displayed perfect drill movements with vigour and zeal. As they marched in unison, the crowd of a couple of thousand spectators, comprising the cadets' friends and families, cheered as they watched their young men make a transition from gentlemen cadets (GCs) to officers.

---

14 Conducted by the Union Public Service Commission for recruitment of Commissioned Officers in the Indian Military Academy, Officers Training Academy, Indian Naval Academy and Indian Air Force Academy.

Also standing in the crowd was the family of GC Saurabh Singh Shekhawat, who had achieved the first milestone of being commissioned into the Indian Army as a second lieutenant, after a year spent overcoming the mental and physical challenges of a demanding training regimen. There were times when Shekhawat had rolled, climbed ropes, performed PT (physical training) and drill, skipped sleep and undergone various other punishments, along with his course mates in the IMA, in his quest to become an officer in Indian Army.

I asked him about his Academy days and what he remembered the most about it. Wistfully, Brig Shekhawat recounted, 'Well, I remember whistles a lot. There would be whistles for every occasion—the drill ustad's whistle, the PT ustad's whistles, and more. Then there were the choicest of scoldings and punishments meted out for the slightest of mistakes. Our seniors and instructors would probe to find faults in us, but we enjoyed it. Front rolling, side rolling, maharaja, running,[15] playing and surviving every moment with your course mates was an absolute delight. I would say those were the best days of my life.'

I asked Maj Gaurav Arya (retd) about the bonds he and Brig Shekhawat shared as course mates. He smiled from ear to ear and told me, 'Ma'am, Shekhu is from IMA, Dehradun, and I am from OTA [Officers Training Academy, Chennai]. But we did parallel courses, and we have cleared the same SSB [Services Selection Board]. It was in the YOs [Young Officers course] that we were finally together. He is very special to me and so my telling of this story is very personal. I have known Shekhawat for the past twenty-eight years. It is not just that he is the winner of multiple gallantry awards, or that he had a brilliant future ahead of him. I have witnessed his integrity and sense

---

15 Various types of punishments given to young gentlemen cadets in IMA to make them mentally and physically tough.

of ethics. Such men are rare today. If you think about it, this man has often put his career at stake for what he believed was right. *Usko farak nahi padta ki general bane na bane, but usko ye fark padta hai ki sham ko can he look himself in the mirror* [He didn't care whether he would make it to the rank of a general or not. He just thought of whether he would be able to face himself in the mirror at the end of the day]. Shekhawat is only accountable to himself.'

The happy conversation with Maj Arya over a hot cup of coffee and sandwiches at a swanky restaurant ended on a humorous note when he told me about how 'Shekhu' was crazy about fitness and kept telling his course mates, including Arya himself, about how they must exercise: 'Ma'am, Shekhawat wants everyone to be at the Shekhawat level of fit—he does not realize that that is difficult for simple human beings like us. I keep telling him, '*Tu SF ka hai, hum nahi* (You are from Special Forces, not us).'

~

July 1995
Probation Period
Special Forces Training School
Himachal Pradesh

Second Lt Shekhawat took his first steps towards the coveted world of the maroon beret, the Balidan Badge and the SF when he was commissioned into 17 Maratha Light Infantry. A small tenure in 17 turned out to be the turning point in his life. His CO of that time, Col Clement Samuel, allowed him to go for SF Probation. Though unplanned, this move turned out to be the best decision of his life.

Brig Shekhawat shared with me, 'I heard in the Academy that Indian paratroopers are the most combat-rich and battle-hardened soldiers in the world. The Special Forces are legendary. I belong to a

Rajputi Shekhawat village, where people in every house have stories of unimaginable courage to tell. I also have an instinctive spirit of adventure, which translates into a strong desire to test myself under the most difficult of circumstances. I thought no other arm of the Army other than the Special Forces was going to provide me that sense of pride and satisfaction, so I volunteered for the Special Forces training as a young officer from 17 Maratha.'

The very first lesson that young Saurabh Singh Shekhawat learned was that an officer does not subscribe to any caste or religion—he usually follows what the boys that he is privileged to command follow. Ustad Havildar Hanuman Ram taught him this on his very first day. When Second Lt Shekhawat reached the Special Forces Training School (SFTS),[16] he faced the ustad and other trainees.

The havildar looked pleased and asked him sweetly, 'How are you?'

Shekhawat replied happily, 'Fine, ustad.'

The havildar replied, 'Fine? Oh, very good, very good! So how was the journey?'

Shekhawat replied enthusiastically, 'Very nice, ustad.'

The havildar seemed intent on creating some kind of permanent bond with the new probationer and asked, even more pleasantly, 'Nice? Very good, very good. So, what is your religion and caste?'

---

16 A training establishment where basic, special and advanced combat training is imparted for executing a wide variety of overt and covert strategic and operational tasks in war and low-intensity conflict. More info at

https://indianarmy.nic.in/Site/FormTemplete/frmTemp1PLargeTC1C. aspx?MnId=xHP+GITkVApjuKpQe1Ttqg==&ParentID=gLaQ5Omh PHHqSli98YqUWA==

Second Lt Shekhawat, for once, was confused. Until now, he had thought that his name itself would indicate his caste. But he played along, replying, 'Ustad, I am a Hindu Rajput from Rajasthan.'

The Havildar flashed the sweetest of smiles and said, 'Accha? [Really?] Hindu Rajput? Okay! Do you see that pond over there? Those are pure waters. Go take a dip.'

Second Lt Shekhawat looked in the direction of the pond. It was basically a pit filled with sewage water and mud. Even as Shekhawat registered the dirt and the foul smell, he dived in, as instructed, for a dip. He was panting when he reported back to the havildar, who smiled at him and said, 'Our conversation is still incomplete. What were you saying? What was your religion and caste?'

Again, Shekhawat replied, 'Hindu Rajput.'

In response, he was asked to roll around on the ground before taking a dip. This continued until it slowly dawned on Shekhawat that he was probably making some kind of mistake to be relentlessly punished like this. Just then, he spied a hoarding on the grounds. On it were printed the words: *Commando Country, Serve with Pride!*

Something clicked in his brain—he finally had a eureka moment.

When the havildar next demanded to know Second Lt Shekhawat's religion and caste, he responded sharply, 'Ustad, my religion is the Special Forces. My caste is the Special Forces.'

The ustad beamed and said, 'Now you are correct! See, I told you, those are pure waters—they immediately open the brain.'

In this way, the ustad gave the young officer an important lesson on religion and caste in the Indian Army. He told him, 'You are an officer—that too, a Special Forces officer. Your religion is the religion of your boys. If your boy is a Hindu, you are a Hindu. If your boy is a Muslim, you are a Muslim. If your boy is a Sikh, you are a Sikh. And if your boy is a Christian, you are a Christian. As an officer, you are everything, you are all of these religions.'

That was a valuable lesson, but it was probably the easiest one young Shekhawat would learn during training and probation. The training period threw challenges at him tough enough to break the spirit of any average young man. Second Lt Shekhawat watched many of his colleagues in probation breaking their legs or simply failing to pass the rigorous training that was intended to 'separate the men from the boys'.

Those who failed were dispatched out of SFTS immediately. But Shekhawat had his heart set on wearing the maroon beret, that distinctive headgear of paratroopers across the world. He wanted to be associated with the same battle-studded glory and showcase his bravery in the face of tremendous odds. More importantly, maroon was a colour that spoke of a legendary fraternity.

In his deep, confident voice, Brig Shekhawat told me, 'If the need arises, I can give first aid, IV treatment and administer injections. I can handle any navigation or communication equipment and demolish bridges, buildings and roads. I have also scored the marksman grade in firing. We are also trained in unarmed combat and more. That is the beauty of the probation period of SF training. It challenges your limits. And if you survive, you gain a never-say-die attitude which stays with you forever.'

He leaned back on his leather revolving chair and said, 'Let me tell you about my escape and evasion exercise.'

~

October 1995
Escape and Evasion Exercise
Forests near SFTS

Operational readiness around the clock and the ability to move rapidly are factors that set the SF apart. The successful completion of the escape and evasion exercise is crucial for young Para SF probationers to clear their probation.

In this exercise, young probationers are made to act like prisoners of war. They must prove their mettle by escaping the 'enemy camps' (set up deep inside the training jungle by the ustads themselves). The instructors also set up ambushes[17] at various points in the wilderness to attack and capture unsuspecting trainees, who are then sent to fugitive camps. From here, the trainees are expected to escape like fugitives, covering a distance of 100 kilometres within a span of twenty-four hours.

Brig Shekhawat laughed out loud as he recollected his experience, saying, 'It rained cats and dogs that day. We were a group of six probationers, and the field-craft exercise turned into a nightmare. The terrain was difficult and there were various obstacles. Many times, we almost drowned in the pits that were filled with rainwater. Our clothes were tattered, and our feet had hundreds of blisters. It felt impossible to take a single step further, but your survival instinct tends to kick-start the adrenaline in you. So, we removed our shoes and hung them around our necks with the help of shoelaces and continued running. We were not given food for several days in the fugitive camps. We also had to evade ambushes. By the time we finished that dreaded escape and evasion exercise and reached our main camp, we looked like Napoleon Bonaparte's defeated army. But since we had managed to finish the exercise in twenty-one hours only, we were also overjoyed. We expected the entire camp to hail our success, as if we had conquered the moon.'

---

17 A long-established military tactic in which combatants take advantage of concealment or the element of surprise to attack unsuspecting enemy combatants from concealed positions, such as behind dense underbrush or hilltops.

He continued, saying, 'In that euphoric high, the men lay down for a while, thinking their torture was finally over. But then the ustad whistled again, signalling another two-and-a-half-kilometre-long BPET[18] right then! We were baffled. Many other probationers protested, but the ustad whistled again. He announced that two minutes had already passed and those who had to pass the exercise now had only six minutes left to do so. So, all the probationers got to their feet and began running again. We completed the drill and waited anxiously in case another shock awaited us. But this time, thankfully, the day was over.'

He smiled and said, 'The idea is to break your spirit—only to rebuild and solidify it with the belief that nothing is impossible for a Special Forces operative.'

Second Lt S.S. Shekhawat's probation finished in October 1995. He then joined 21 Para (SF), having been converted from 21 Maratha Light Infantry as a part of the extension of the SF battlions.

~

Dec 1995–March 1996
High Altitude Warfare School
Gulmarg
Kashmir

The young boy who came from the hot, sandy and arid land of Rajasthan never imagined that he would one day be exploring steep, gigantic snow-laden icy peaks. But, as they say, you don't choose the mountains, the mountains choose you. Shekhawat's path was destined. He was nominated to undertake a mountaineering course—a fluke which occurred because he was asked to take the

---

18 The Battle Physical Efficiency Test (BPET) includes running within a limited frame of time to pass the test.

place of a senior who, for some reason, could not be available for the course at the High Altitude Warfare School (HAWS) in Gulmarg, Kashmir.[19]

Shekhawat didn't know it then, but he would soon form an everlasting bond with the mountains. Later, he would think of the irony: the desert and the mountains seem very different, yet they are similar. Both are inhospitable, harsh, barren, uninhibited and unsmiling, but are also simultaneously majestic, magnificent, mysterious, and immensely beautiful. A staggering combination of fire and ice, if you will!

When Shekhawat began his journey at HAWS, he knew next to nothing about mountains. He had even unwittingly brought wrong-sized shoes, which caused him great difficulty while climbing. As a result, excruciating blisters developed on his feet. However, as he acclimatized, the blisters began to heal, and he began to exhibit the signs of a natural mountaineer. He showed a natural affinity for the terrain, notorious for its freezing, howling winds and sub-zero temperatures, which punished even the most hardened mountaineers with frostbite. A single false step could mean certain death.

It was at HAWS that Shekhawat learned all about high-altitude mountain warfare—from understanding how to survive at high altitudes and snowbound areas in the winter months to night

---

19 A training establishment set up in Gulmarg, Kashmir. The training imparted consisted mainly of skiing techniques, mountain lore, and patrolling on skis and for specialized training and dissemination of approved doctrines in mountain, high altitude and snow warfare.

More info on high altitude mountain warfare at https://indianarmy. nic.in/Site/FormTemplete/frmTemp1PLargeTC1C.aspx?MnId= Y0t+SBnaVl/ny0VKQA/L5A==&ParentID=994zUExwEnwXtWoOe ylONg==

operations in the snow. He skied with a load, made igloos and mastered various climbing techniques.

Still, even though he seemed to be a natural, the training period was not easy. Shekhawat's body would be sore at the end of the day, his feet numb despite the layers of extra socks. Blisters had an annoying habit of sneaking back into life. Many times, the training took a severe toll on his spirits and sanity. But he held on, remembering his fiancée, Renuka Rathore. On long and lonely nights, after the brutal training, Shekhawat kept his heart warm with memories of Renuka. The two had got engaged in 1994, but their marriage had to be postponed because of his training courses and schedules. But the young couple kept their romance alive by writing long letters to each other. Those letters and the memories of the happy times they had had together kept Shekhawat going even when he felt like giving up.

~

Marriage and Family Life
Jodhpur
Rajasthan

Brig Saurabh Singh Shekhawat's mother, Dr Shraddha Chauhan,[20] is an eminent scholar, with a PhD and DLit in Vedic literature. She was a Sanskrit professor at the University of Jodhpur. One of her students, Renuka Rathore, had caught her eye when she had been on the lookout for a good girl for her son to marry. After consulting with her husband, Dr Jaswant Singh Shekhawat, Dr Chauhan had approached Renuka's family.

The Rathores were confused. Renuka was still studying, and they wanted her to continue on this path for some more time. But Renuka

---

20 Women marrying into the Shekhawat family retain their maiden names, while the offspring take the surname of their fathers.

had different ideas. She knew Saurabh already. They had had a few chance encounters when Renuka had been doing her PhD under the supervision of Saurabh's maternal grandfather, Dr Fatah Singh, who lived with Saurabh's family in Jodhpur.[21] Renuka visited their house often in those days and had even stayed at their house on occasion. Saurabh had seen her around now and then. They had locked eyes with each other, and shy smiles had been exchanged. But since they belonged to conservative Rajput families, neither had made the first move and initiated conversation. It was only when Dr Chauhan entered the picture that things began to move ahead.

Dr Shraddha told me, 'Renuka was my student, and I had observed her for a long time. She belonged to a good family. She was academically brilliant, and I wanted her to study more. I asked Saurabh if he would be willing to marry her. He said though he liked the girl, his job was risky and he would be away most of the time. So, if he got married, we would have to look after his family too. We readily agreed.'

And so it came to be that Renuka Rathore entered Saurabh Singh Shekhawat's action- and adventure-filled life. Renuka is a simple, sweet and very accomplished woman. The couple got engaged on 21 May 1994. But actual marriage remained elusive—Shekhawat was always away at some training centre or the other.

Renuka Rathore remembers those days with great nostalgia. She told me, 'Swapnil, I would say those two-and-a-half years of our courtship were more romantic than the last twenty-five years of our marital life. I remember how we used to write long letters and waited endlessly to meet each other. After our marriage, we hardly stayed together. While I was occupied managing the home, he was busy serving the nation—I respected that.'

---

21 Dr Shraddha Chauhan and Dr Jaswant Singh Shekhawat lived in jodhpur from the 1980s to the 2000s.

Eventually, as soon as Shekhawat completed his mountaineering courses, he returned home. He and Renuka were married on 25 November 1996 in a simple Rajput wedding in Jodhpur at the bride's home. The residents of several villages around the area attended the wedding ceremony, which was conducted according to Vedic rituals.

Sadly, there was no time to extend the festivities. Within ten days of the wedding, Lt Saurabh Singh Shekhawat went away for the commando training course, leaving behind his new bride. It was the first of many long periods of being apart in their marriage. Gradually, waiting to be with each other became an integral part of their lives.

Brig Shekhawat told me, 'In the twenty-five years of our marriage, we have hardly been together in one place because of my job profile and other commitments. If not participating in operations in the rural hinterland, I could otherwise be found summiting mountains or training for new skills. I was always on the move—UN postings, courses and missions. All this kept us apart. Eventually, we gave up on our dreams of staying together and leading a regular family life.'

Such is the harsh lived reality of the families of SF operatives. Many a time, they struggle to stay together as a couple, as one family. It wouldn't be wrong to say that the family of a fauji serves the nation as much as the fauji himself.

Renuka Rathore told me fondly, 'There were periods when Saurabh did not return home for several months. I would see other children demanding things from their fathers, while my girls craved for his mere physical presence. That was tough. I remember doing all the household chores by myself, from sweeping to cooking to washing the clothes. I had no help, as Saurabh did not believe in extending fauji resources to the family. I also had a full-time job as an assistant professor at the University of Jodhpur. Despite all this, I managed and raised my daughters well. This taught them to be independent and self-sufficient. Now that they are studying abroad using their own funds, they thank me for teaching them the art of self-sufficiency early in their lives.'

After their girls were born, Shekhawat, now a Major, decided to construct a house, which he fondly named 'Sagarmatha', the Nepali name for Everest, the great mountain he had scaled three times. The home provided stability for his family, allowing Renuka to focus on her career and his girls to pursue their education, unlike most military families who move from station to station across the country all the time.

Today, the women of Sagarmatha and Brig Shekhawat are equally proud of each other's achievements. Eyes gleaming, with a quiver of pride in his voice, he says, 'I give the entire credit of raising our girls so wonderfully to my wife, Renuka. She had the strength and tenacity to bring up two strong women while staying on top of her own career. Whoever I am today, or whatever my girls have achieved, it is all because of my wife and the support of my parents.'

Then, laughing, he adds, 'Though, ma'am, I would certainly like to give myself the credit of teaching my girls the art of horseback riding, mountaineering and trekking. I have also taken them on several difficult treks. This instilled a go-getter attitude in them—they don't give up easily. My girls are my pride.'

Brig S.S. Shekhawat's parents have indeed always been a great support. In particular, his mother, Dr Shraddha Chauhan, has been a towering pillar in this regard. His father, Dr Jaswant Singh Shekhawat, has a PhD in horticulture and retired as professor of horticulture at Rajasthan Agricultural University. He also has a great interest in Indian culture and heritage and spends his nights reading history books and researching various topics.

Maj Neetu Chandran, an officer who has served under Brig Shekhawat and has had the privilege of meeting his family, told me, 'One look at his family and his roots and you know where his legacy comes from and what pushes him to achieve the impossible. When you speak to his father, his mother or his wife, you are dumbfounded by their grace, elegance and, most importantly, their vast knowledge.'

I agree with Maj Chandran. The warmth of the Shekhawat family touched my heart in ways that I didn't expect, as did their visible pride in Brig Saurabh Singh Shekhawat's accomplishments. No wonder giving up has never been an option for him.

~

March–May 2001
8,848 metres above sea level, Mount Everest

Mountaineering expeditions are planned almost as intricately as complex attack operations. There are many factors to contend with: high altitudes, low atmospheric pressure, rugged ridges of ice and rock, deep crevasses and deadly storms. Perhaps this is why SF operatives have distinguished themselves time and time again by scaling summits.

Brig Saurabh Singh Shekhawat is one of a handful of Indian Army officers to summit multiple peaks, and perhaps the only one to summit twenty-one peaks. This includes Mount Everest, which he summited thrice in 2001 as part of the Indian Army Everest expedition team; then in 2003, as part of the Indo-Nepal Army Everest Massif Expedition, where he was the deputy climbing leader of the expedition; and finally in 2005, as the leader of the first-ever women's expedition of the Indian Army. He has also summited Mount Kilimanjaro (the highest peak in Africa), Mont Blanc (the highest peak in the Alps) and Mount Nun (a mountain massif of the greater Himalayan range), Mount Ruwenzori in Congo, Marble wall peak in Kazakhstan and many 6,000- and 7,000-metre peaks in India and Nepal, to name a few others.[22]

---

22 In June 2022, when I was editing this book, he conquered two virgin peaks in Sub Sector North on the Line of Actual Control: Shahi

These feats are extraordinary. In 2021, when I was writing this book, I found that very few officers had accomplished what Brig Shekhawat managed nearly fifteen years ago. There has been no sensational coverage, in the media or otherwise, on his achievements. Yet, he has earned a place for himself in the annals of Indian mountaineering history.

Understandably, summiting Mount Everest has been one of Saurabh Singh Shekhawat's most exhilarating experiences to date. When I asked him about this extraordinary achievement, he said modestly, 'Everest is beautiful but also dangerous. She is both kind and cruel. On the one hand, she allows even the most novice climbers, from the blind to the disabled, to reach her peak. And on the other hand, even the most legendary climbers have sometimes been unable to conquer her. She has kept many world-record holders hanging on for dear life on her slopes. It is only at her mercy that you are allowed to scale her and come back safely. So, I won't say that I have *conquered* Everest. I will say that I have *survived* her.'

In 2001, Shekhawat was part of a twenty-four-member[23] Indian Army Everest expedition team. He was selected out of 180 Army personnel who volunteered for the ambitious team determined to summit Everest after the 1985 expedition disaster,[24] where five Indian Army officers had been killed.

I want to highlight the history of Indian Army's Mount Everest expeditions a bit to the readers here to highlight the importance of Brig Shekhawat's Mount Everest Expeditions for which he has

---

Kangri (6,934 metres) and Silver Peak (6,871 metres). For pictures and reference, please see: https://twitter.com/ firefurycorps/status/1543423 287463936000?lang=en

23 Of the twenty-four-member team, twelve were climbers while the rest were part of the support team.

24 Suman Dubey, 'Indian mountaineering suffers worst disaster as five army officers die on Everest', *India Today*, 15 November 1985.

been recipient of third highest peacetime gallantry award in India, Shaurya Chakra, and also of the distinguished service medal Vishisht Seva Medal for another Mount Everest expedition in 2005. Capt Avtar Singh Cheema from 7 Para was the first Indian man and the sixteenth person in the world to climb Mount Everest as part of the Indian Mount Everest expedition team in 1965. The Indian Army launched its first Everest expedition in 1985, which turned out to be a disaster.[25] After 16 years in 2001, the Indian Army decided to send another Everest expedition team, and the pressure was enormous. Rigorous tests were undertaken to select the team members under the leadership of Col Krishnan Kumar. The 24-member team was flagged off by the COAS General Sundararajan Padmanabhan on 1 Mar 2001, which eventually successfully completed the mission against all odds.[26]

When I asked about his selection, Brig Shekhawat says, 'As a Para SF officer, I had already completed my basic and advanced winter warfare course from HAWS, and that learning helped in my selection as the youngest member of this ambitious team.'

The team reached the Everest base camp on 3 April after a monthlong trek through Nepal—this helped the team to properly acclimatize to the weather while also allowing them to explore the countryside. Capt Shekhawat reached the base camp as an advance party commander and set up a big camp with ten palatial Arctic medium tents, larger in comparison to the usual dome tents used by most others. These Arctic tents ended up becoming a major stomping

---

25 https://www.upi.com/Archives/1985/10/12/Indian-climber-left-to-die-on-Everest/1180497937600/

26 https://indianarmy.nic.in/Site/FormTemplete/frmTempSimple.asp x?MnId=6PsOUpA6qzPyj308E0ObWQ==&ParentID=kIs9WucW Wb8NPwRWcd5xHQ==#:~:text=The%20successful%20Army%20 expedition%20to,COAS%20on%202001%20Mar%202001 (Mt Everest (8848M) Expedition 2001)

ground for climbers from all over the world—a place where they could enjoy free coffee, a rare luxury in the tightly rationed and highly expensive Everest base camp.

While the Indian Army team rapidly prepped for the treacherous climbing that lay ahead, Capt Shekhawat learnt the first of many lessons Everest would teach him. His adrenaline was always pumping at the time, leading him to compete with the sherpas, loading ferry and marching up the slopes at the same speed as them—a grave mistake. The sherpas are the champions of Everest. They are born and brought up in this terrain. An outsider can never compete with them.

Capt Shekhawat's mistake resulted in a severe upper respiratory tract infection. He developed fever, bleeding gums and other high-altitude-related complications. He was eventually evacuated to the Nepal army hospital in Kathmandu via helicopter. As he watched the steep peaks of the Himalayas disappearing out of sight beneath the aircraft, the Captain's heart sank.

The mountains were Shekhawat's calling. Summiting Everest had been his dream. Now, it felt like that the dream was out of his reach. Once you are evacuated in the middle of the expedition, over medical issues, the chances are high that you will be unable to rejoin your peers. That is what happened. Shekhawat was in hospital for four days. He asked to return to the base camp as soon as he felt better but his requests were denied.

He was restless. Giving up had never been an option. So, he lied to the doctors about needing to visit the Indian Embassy in Kathmandu. As soon as he was duly discharged, he immediately started trekking back to the base camp. Col K. Kumar was surprised to see him there. When Shekhawat insisted on climbing, saying that he was absolutely fine, Kumar allowed him to rejoin the group. However, he left him with a great piece of advice, which Brig Shekhawat has followed ever since, every time he climbs a mountain: 'Don't be a gama in the land of the lamas. Acclimatize, then climb. The slower you climb, the higher you reach and the faster you climb, the lower you reach."

The Indian Army team was determined to summit Mount Everest. It was a matter of great pride to them that they had been chosen to fulfil the dreams of the team before them, which had met with such an unfortunate end in 1985. The team first undertook the treacherous crossing of the Khumbu Glacier[27] in ten days before they reached the South Col[28] on 10 May 2001. Unfortunately, they were forced to return to base camp due to bad weather. Finally, after a few days, when the weather improved, on 19 May 2001, the team set out again to reach their final summit. Again, they returned after the route was buried in a sudden snowstorm.

In all this action, the enthusiastic Capt Shekhawat was at the forefront of opening the routes, putting up ladders along with the five other sherpas responsible for the task. It took them three more days to reopen the route after the snowstorm. It was not just a great learning experience but also a liberating and joyous one to work alongside his comrades.

When the Indian Army team reached the South Col, popularly known as Death Zone, their determination to summit the peak only strengthened. The peak was only 1.74 kilometres away from the South Col, but it would have taken them nine or ten hours to complete the journey.

Winds raged at 15 to 35 knots (27.78–64.82 km/hr) at the South Col, and temperatures often plummeted to as low as -50 degrees

---

27 The world's highest glacier.

28 A a sharp-edged col (lowest point of mountain ridge) between Mount Everest and Lhotse, the highest and fourth-highest mountains in the world respectively. The South Col is typically swept by high winds. When climbers attempt to climb Everest from the southeast ridge in Nepal, their final camp (usually Camp IV) is situated on the South Col

Celsius. The rarefied air at the altitude of 8,000 metres made simple acts like boiling water for tea or the tying of bootlaces painfully laborious. Nights of fitful and troubled sleep were also one of the outcomes of trying to summit the Sagarmatha, the goddess of the sky.

But just reaching and surviving the Death Zone filled Capt Shekhawat's heart with excitement. The great mountain challenged his every limit. Was this not what Capt Shekhawat had craved all his life?

Things were different in 2001. In Jodhpur, when people heard about Dr Shraddha Chauhan's son undertaking such an arduous and audacious trek, they were astonished. They asked her how she could have let him to go on a potentially life-threatening quest. Dr Chauhan's response was always staunchly proud: 'Why would I be scared? I stopped getting scared when he joined the fauj. I gave my son to the nation. We all are surrounded by death—and it can approach one even when one is sleeping, so why should one be afraid of death? My son has always been fearless, and I am proud that he is trying to achieve something great.'

Finally, D-Day arrived, and on 22 May 2001, the final nightlong quest to the summit began. Shekhawat's team,[29] which comprised seven members, junior commissioned officers, non-commissioned officers and three sherpas, reached Everest's summit on 23 May 2001. After opening routes and fixing ropes in the blustery weather above the South Summit[30] and the Hillary Step[31], they succeeded

---

29 The twelve-member climbing party was divided into two teams. Capt Shekhawat's team comprised seven men and climbed before the other team.

30 A subsidiary peak to the Mount Everest peak. Its elevation is higher than the second-highest mountain on earth, K2, but it is not considered a separate mountain as its prominence is only 11 metres.

31 A nearly vertical rock face located near the summit of Mount Everest. Located on the southeast ridge, halfway between the South Summit and

in summiting the Everest, braving severe winds and subfreezing conditions.

Shekhawat's beard was powdered with snow. His throat was parched; it felt a little like someone had raked nails over it. But his jubilation knew no bounds, especially when he took out the national flag and the flag of the Para Special Forces and hoisted them at the top of the world.

On his way down, Shekhawat called Renuka on a satellite phone. He knew his wife had been praying and fasting for his successful summit. He had to call her and let her know.

'Renuka, I summited Everest. I have done it. I still cannot believe it,' he said.

There was joy, jubilation and pride in his voice. After a two-second pause, his wife asked calmly, 'It is okay, but tell me, from where are you calling?'

Capt Shekhawat replied, 'Well, from the South Summit itself!'

Renuka was worried sick for several days. Her husband may have created some new climbing record, but for her, his safety was paramount. As the wife of a mountaineering enthusiast, she knew all about the majesty, myths and perils surrounding Everest. So, she said simply to him, 'Oh! Don't be so happy. First, climb down safely, and then call'—and hung up.

Recalling the moment, Brig Shekhawat laughed out loud. He told me, 'Look, ma'am, I had just summited Everest. It was a huge feat. The news had already spread. I was receiving congratulatory notes from all across the world, but my wife was not at all impressed.'

I smiled. Even if you are the bravest man alive, it would always be the wife who reins over the kingdom, I suppose.

---

the true summit, the Hillary Step is the most technically difficult part of the typical Nepal-side Everest climb.

In this team of twelve climbers, ten had completed the summit.[32] The extraordinary achievement by the Indian Army brought much glory to the nation. It broke the jinx of the Indian Army's inability to summit Everest, since the last unsuccessful attempt in 1985, and encouraged many from the Army, Navy and Air Force to attempt to summit this glorious, cruel mountain. In fact, the Postal Services Department even issued commemorative postage stamps honouring this 2001 expedition.[33]

Later, the then President of India, Dr A.P.J. Abdul Kalam, awarded Capt S.S. Shekhawat the Shaurya Chakra for his undaunting spirit and gallantry in the face of high-risk weather and terrain extremities.

However, that was not the end of Brigadier S.S. Shekhawat's quest to challenge his endurance and test his faith. He conquered the summit of Everest thrice, and each time the challenges remained unpredictable and the terrain formidable. He related to me how painful it was to lose his fellow climber and comrade Havildar Bhim Bahadur Bhujel during the 2003 Indo–Nepal Army joint expedition,[34] when he was the deputy leader of the team.

Brig S.S. Shekhawat also has fond memories of leading the Indian Army's first women expedition team, which created history by summiting Everest in 2005 under his leadership. The team comprised six women officers from the Army, two girl cadets from the National Cadet Corps and twelve experienced Army climbers. On 2 June 2005,

---

32 The twelve-member part of climbers was divided into two parties. Two of the climbers could not reach the peak.

33 https://istampgallery.com/indian-army-everest-expedition-2001/

34 The Ascent of Everest and Lhotse 2003, Indian Mountaineering Foundation, http://bit.ly/3xMjTkD

the team summited Mount Everest from the Chinese-occupied Tibet side, through the North Col[35] route.

But this victory did not come easily. Hard work, determination and outstanding leadership made this dream possible.

When Shekhawat was appointed the team leader, he immediately took the twenty women to Siachen Battle School for further training. The girls compare him to Shah Rukh Khan's character in the 2007 Hindi-language sports drama *Chak De! India*. He also turned out to be a tough taskmaster who united a pool of talented girls and put together a team that made the impossible possible.

Shekhawat was strict with the women—there were a lot of physical hurdles and weight training. He made them run for 8–10 kilometres daily and also trained them to climb with 15-kilogram rucksacks on their backs. Fractures and injuries became part of their daily lives. Sometimes, the girls would give up, but no excuses were entertained, nor were any concessions provided.

He told them, 'Everest is unforgiving; if you hope to get to the summit, you need to toughen up. The mountain does not discriminate between the sexes, nor does she give allowances. The circumstances will be the same for you as they are for anyone else, and your survival is at stake. It does not matter to me if all of you remain at the base camp—if I take even one lady there, she will need to be the fittest.'

Shekhawat kept pushing their limits and, eventually, the girls recorded their names in the annals of history by becoming the first Indian Army women's team to summit the dangerously beautiful

---

35 Referred to as Chang la in Tibetan, it is the sharp-edged pass carved by glaciers in the ridge connecting Mount Everest and Changtse in Tibet. Before 1950, most Everest expeditions went from Tibet and via the North Col Now most expeditions go from Nepal via the South Col

Mount Everest. The team was later felicitated by Dr A.P.J. Abdul Kalam. All the family members of the climbers were present at Rashtrapati Bhavan, including Shekhawat's little girls, who watched everything with astonishment and refused to leave their father's arms.

In his speech, Dr Kalam said, 'The very basic core of a man's living spirit is his passion for adventure. The joy of life comes out of our experiences with new encounters. The spirit of adventure should be alive in the collective conscience of humankind. It is what keeps you moving. This spirit of adventure has taken our race to the moon and Mars. These endeavours set the bar for human ethics and expectations. I am very proud of these women who made the impossible the possible.'

Maj Saurabh Singh Shekhawat was later awarded the Vishisht Seva Medal for his selfless and distinguished service of a high order to the nation.

He has summited twenty-one more peaks since then. Recently, on 3 July 2022, the Twitter handle of the Indian Army's Fire and Fury Corps announced fifty-one-year-old Brig Shekhawat's most recent expedition of the challenging and till then unclimbed twin peaks of Mt Shahi Kangri (6,934 metres) and Silver Peak (6,871 metres) in the hostile and famous Sub Sector North on the Line of Actual Control.

Brig Saurabh Singh Shekhawat has kept seeking out new horizons and challenging the impossible with every task he undertakes—from spine-chillingly dangerous operations in the jungles of the Northeast to diving off an aircraft with a full weapon load at 25,000 feet in the sky.

The man, I thought to myself, was unstoppable.

~

1997
The Northeast

Brig Saurabh Singh Shekhawat is one of the highest decorated serving officers of the Indian Army, with an outstanding operational profile. He has participated in almost all ongoing operations of the Indian Army: Operation Hifazat in the Northeast, Operation Parakram in Kashmir, Operation Rakshak in Kashmir, Operation Vijay in Kargil, Operation Rhino in Assam, Operation Orchid in Nagaland and Operation Meghdoot in the Siachen Glacier.

In the Northeast, as a young 21 Para Special Forces operative, he participated in his first operation in Assam. The young officer had just completed his Para probation. He did not feel any sense of euphoria when he took his first shot, nor was he nervous when bullets flew around him. This boy, belonging to the hot and harsh lands of Rajasthan, who had previously carried hundreds of sacks on his back, who had run behind camels and who had worked hard on difficult, arid fields, had an inherent capacity to calmly deal with violence, blood and gore.

Armed with these robust qualities, it was certainly an odd quirk for Shekhawat, then, to be fascinated by astrology. During his first few operations, he had checked the palms of several dead terrorists, a practice he continued for quite some time. More often than not, he told me, he discovered long, flowing lifelines on the palms of the deadliest dead terrorists.

'So, you see, I stopped believing in astrology since then,' he explains to me. 'Life is unpredictable at the end of the day, and only the supreme forces decide who will live or leave for their heavenly abode. One must not lead one's life in fear of death. It will come when it has to come. But, yes, what matters is how you chose to lead your life, and whether it was on your own terms.'

Shekhawat has also faced times when his heart dealt with excruciating pain, when his vision blurred with tears and when his spirit sagged. The death of dear friends or loyal comrades, felled by bullets or blasts, have broken his heart several times.

There were also terrible moments with happy endings, such as his rescue of a three-year-old girl during one of his operations. Shekhawat had spotted her in a pool of blood, lying among the dead—a bullet had injured her leg. He rushed her to hospital, holding her in his arms. Thankfully, she survived.

~

September 1999
Somewhere in north Kashmir

There was word of heavy infiltration by a large group of foreign terrorists in an area in north Kashmir. When the Army got wind of this, ambushes were laid at two places, on higher grounds, on tracks frequented by the terrorists. Capt Shekhawat, a troop commander, occupied a hillock in the area, while his senior team commander, Maj Deependra Singh Sengar (retd),[36] occupied the target location at another hillock, 4 kilometres away from Shekhawat's location.

After two days of waiting in ambush, there was still no contact with the terrorists. However, on the third night, Maj Sengar's team spotted the terrorists, and a brief firefight took place. There were shouts and shrieks, but since the encounter took place in the thick of the night, nothing was visible. Still, Maj Sengar was sure that some terrorists had been hit by bullets. He immediately called Capt

36 https://www.hindustantimes.com/india-news/vijay-diwas-special-major-deep-singh-sengar-who-defied-death-twice-recalls-fighting-the-kargil-war-101627226310112.html; https://starsunfolded.com/major-deependra-sengar/

Shekhawat, who was at another location, and said, 'Shekhu, it looks like some of the terrorists have been hit. I have no idea if any of them have been killed, but some have certainly been injured. Send tracker dogs tomorrow morning. They will lead us to the rest of the party.'

The next day, as soon as dawn broke, Capt Shekhawat requested a dog squad from the nearby Rashtriya Rifles unit. As soon as the dog squad sniffed the blood, they started running over the blood trail. Maj Sengar and his squad of six men ran behind them. The terrorists had been providing first aid to their own injured men when they heard the tracker dogs coming for them. Instantly, they hid in the thick vegetation. As soon as Maj Sengar's squad neared, a firefight broke out. Many were hit, including Sengar himself, who immediately went on air and contacted his troop's commander, Capt Shekhawat.

'Sengar for Shekhu, Sengar for Shekhu,' echoed the secure-communication sets.

Capt Shekhawat responded, 'Shekhu for Sengar, Shekhu for Sengar.'

Maj Sengar said, 'I have been hit. I don't know if I will survive or not, but these terrorists should not escape. You come down and take position. I am passing you the coordinates.'

There was no time for planning. Capt Shekhawat immediately scrapped his ambush and ran with his squad of six men towards the given coordinates. He heard heavy firing from both sides in the middle of the jungle when he reached the place. The thick vegetation made it almost impossible to see where his team was and where the terrorists were located. At one point, he also received a radio call from Maj Sengar's buddy sepoy, Dut Ram Ghale,[37] who said, 'Saab, come soon, they are too many—and I am the only one left here with

---

[37] https://static.pib.gov.in/WriteReadData/specificdocs/documents/2022/jan/doc20221258601.pdf

Sengar saab. I am firing and warding them off, but I don't know how long it would be possible. They will be upon me at any moment.'

These are the moments when life is measured in a matter of seconds. Capt Shekhawat asked Ghale to take single shots to be sure of his direction and locate the terrorists quickly. As soon as Ghale started taking single shots, Shekhawat was clear about the location of the terrorists. He immediately asked his boys to fire an 84 mm Carl Gustav rocket launcher in their direction. Eight rocket-launcher rounds had to be fired to break their ambush and cause the terrorists to flee. The action provided Capt Shekhawat with a window of time to reach Maj Sengar's position and evacuate him. As Capt Shekhawat's buddy moved ahead to pick up a fallen gun, he was shot at. The bullet went wide, but it alerted them to the presence of a sniper. Shekhawat and his men dived for cover as the sniper continued to shoot out of the thickly forested area.

There was no way to ascertain where the sniper was hiding, so Ghale strategically began abusing him loudly. After a few moments, the terrorist got agitated and shouted, *'Gaali kyu de raha hai? Gaali mat de* (Why are you swearing? Stop it)!'

As soon as they had some idea of the sniper's position, Capt Shekhawat took the reins and engaged him in conversation, saying, *'Mujahid, kaha ka rehne wala hai* (Mujahid, which place do you belong to)?'

The terrorist realized he was talking to an officer and answered respectfully, *'Janab, Pakistani hu* (Sir, I am a Pakistani)'

Capt Shekhawat shouted again, *'Yaha khoon kharaba kyu kar raha hai? Mara jayega tu* (Why are you on a killing spree here? You will be killed)!'

The mujahid replied, *'Yeh to Allah Subhanahu wa ta'ala ka hukum hai. Hum to jihad karte hai. Aapke aur mere samjhane ki baat nahi hai, hum dono to sirf pyade hai* (It is ordained by Allah Subhanahu wa ta'ala. We fight on His behalf. It's not something that can be understood by either you or me. I and you both are only foot soldiers).'

By thus engaging the sniper in conversation, Capt Shekhawat had a pretty clear idea of his location by that time. He tossed a grenade in his direction; the terrorist responded with a hail of bullets. Unpinning yet another grenade, Shekhawat aimed again. This time, the firing stopped. There was a lull in the air.

Wanting to find out whether the sniper was alive or not, Capt Shekhawat asked the terrorists hidden well under the canopy of trees, '*Mujahid, lagi toh nahi* (Mujahid, are you hit)?'

The terrorist replied calmly, '*Ji janab, aapne jo grenade fenka, usi se lagi hai. Raat me bhi goli lagi hai mujhe janab* (Yes, sir, the grenade you threw has hit me. I also have a bullet wound from last night's fight).'

Capt Shekhawat, trying to get him to surrender, said, '*Toh bas ab tu mar jayega. Abhi bhi time hai—surrender kar de, hum ilaj karwa denge* (You may die any second. But you still have time—surrender and we will provide you with medical aid).'

The terrorist replied, '*Janab, marna to hai hi. Aap kaun se reh jaoge. Sabne jana hai ek din* (Sir, we all have to die one day. You will die too).'

Precisely at that moment, Capt Shekhawat's radio operator heard the conversation. In an attempt to take a shot, he started climbing up the hillock, but the injured terrorist still had not lost his spirit and fired at him. The soldier stumbled and fell. Capt Shekhawat shouted his name and heaved a sigh of relief when the operator said, '*Bachh gaya, saab! Goli radio set ko lagi hai* (I was saved, sir! The radio set was hit).'

Just then, the terrorist moved his position. That was all that was needed for the ever-alert Shekhawat to take his shot and prove himself to be the marksman he knew he was. The bullet hit the man dead in the centre of his forehead.

Later, when the news caught the attention of the media and made headlines, one of the slain terrorists was recognized as a high-ranking terrorist commander who was helping the foreign terrorists to infiltrate the area. He had also been responsible for mass-scale recruitment, indoctrinating locals in the Valley and even killing local villagers who dared to defy him.

Brig Shekhawat said, 'We lost three of our boys, which was a terrible cost to pay for a successful operation.'

I remember feeling a deep sense of sadness listening to Brig Shekhawat. Duty demands blood and sacrifice. Perhaps this is why the Balidan badge is worn by a select few who are fearless in heart, mind body and soul.

However, this was not the end of the trail of blood that Shekhawat traversed in the name of performing his duty to the nation. He fought in Kargil in 1999 as well, where, during a cliff-assault mission, he was hit by Pakistani splinters while hiding under a rock. He did not even realize this at the time, and the splinters remained hidden in his body for a long period and were cleansed much later when they turned painful.

As a storyteller, I found it difficult to cover all of Brig S.S. Shekhawat's exploits in a few pages. But as I sipped green tea in his office and nibbled on roasted nuts and cookies baked by the mess cook, I knew that I would have to tell at least some of the stories that needed to be told.

~

May 2007
Somewhere near the India-Myanmar border

There was a small village located on the India–Myanmar border. The line demarcating the two countries runs straight through the village, dividing it so that one half is on India's side while the other is on the Myanmar side.

The village belonged to one group of tribes, but an insurgent group of another tribe killed many of the villagers and pushed others out of that village.

It became a ghost village, occupied only by terrorists, who used it as their base to operate in India and freely cross the border into Myanmar, safe from the eyes of Indian border security and law enforcement agencies.

An informant approached the Indian Army, asking for help. He stated that the village belonged to his tribe and that it had been illegally occupied by the insurgents. He said that it not only disturbed the power equation in the area but also posed a grave threat to the sovereignty of India.

The 21 Para Special Forces decided to help, and Lieutenant Col Shekhawat, a team commander at that point, was asked to lead the operation to liberate the village.

Lt Col Shekhawat asked the informer to lead the way and to the village. Reaching there meant undertaking a three-day march across difficult and rugged terrain, the ground studded with landmines. As a result, Shekhawat and his team had to take extra precautions to keep themselves safe.

Once they reached, Shekhawat asked his team to harbour.[38] In the dense jungle, different squads of his team occupied various positions, using the ground, bushes, fallen trees and thick vegetation to camouflage themselves. It was already late evening, and by the time they finished their task, it was dark. Lt Col Shekhawat planned

---

38 A military practice of occupying tactical positions in order to attain dominance over the area before launching the actual operation and creating a temporary base where soldiers could also fall back for rest, recoup and recharge

activities for the following day with his troop commanders. He asked the informer to show them the village from a vantage point. Shekhawat thought it would be a good idea to see the target area himself before they launched an actual attack.

The following day, even before the commando base had started operating at full capacity, the men heard a noise—someone was cutting wood. Their informer whispered to Shekhawat that these were Burmese citizens who crossed the border every day to cut the bamboo trees that grew on Indian soil. They were supporters of the terrorists, who provided them with firewood and other essentials. The commandos were on alert, ready to fire should the occasion arise. But they continued to wait patiently—with bated breath—for the woodcutters to leave. They knew that should trouble break out, the element of surprise would be their biggest defence. Thankfully, none of the woodcutters noticed them, and they eventually left with lots of wood and bamboo. As soon as they had gone, Lt Col Shekhawat resumed his attempt to recce the village with the help of the informer, and asked the troop commanders to be on alert.

Along with the informer and six other men from his squad, Shekhawat left for the village, located 700 metres from the harbour area. After walking for an hour, the outskirts of the village came into view. Some instinct warned him of danger. Alarmed, he grabbed the informer by the collar, pushing him into a nearby nallah and simultaneously signalling his men to hide as well. He asked through gritted teeth, 'Bloody hell, where are you taking us?'

The informer replied innocently, 'To the village, sir.'

Astonished, Lt Col Shekhawat asked him, 'Why are you taking us straight to the village? It is just seven of us here. How will we fight all of them?'

A little confused, the informer replied meekly, 'I thought you guys were commandos. You are all so brave that I thought just seven of you were enough to fight and eliminate fifty terrorists on your own. Why else would you ask me to show you the village?'

Lt Col Shekhawat let out a sigh. That little misunderstanding could not only have had them killed but also jeopardized their well-planned mission. He hissed under his breath, 'I asked you to show us the village from a safe and suitable viewpoint so that we could establish surveillance and recce in the area before launching the actual operation. And here you are leading us straight into a death trap!'

The informer realized his mistake. He changed course and took them to a vantage point. But even from a height, they could only see one hut because of the thick, dense vegetation in the jungle. This provided an excellent cover to the village and hid it from aerial surveillance.

Shekhawat asked one of the boys from his squad to climb the tallest tree in sight and report back on what he saw via satellite phone. The soldier did so and informed them that he was able to see all the huts from his position. He confirmed the presence of terrorists in the village. Shekhawat ordered him to establish a machan in the tree and report back on the activities of the terrorists. He wanted to know what a regular day looked like for the inhabitants before he took any action. The boy took out his toggle rope, made a seat harness, and with the help of karabiner, secured the harness and himself in it and started observing the village.

Soon, the soldier reported that some terrorists were sitting outside huts with their AK-47s, some were going through some sort of records of files, and some were managing the rations and doing other regular chores. A sentry post was also there. Shekhawat made a mental note to neutralize the sentry, first and foremost, during the attack.

After noting the activities of the terrorists through the day, the squad returned to their base by evening. But there was no time to rest. Shekhawat immediately called a meeting of all his officers and briefed them on what had been observed during the day. The encounter plan would have to be revised in light of these new inputs.

Shekhawat strategically established several 'stop and cut off' groups whose duty was to stall and kill any militants who might break the cordon and run through. Then, he established a support group, which was responsible for using rocket launchers or suppressing fire if and when the main assault party moved towards their target. It was meant to divert the terrorists. Then came the firebase party[39] and the main raiding party,[40] known as the 'assault group', led by Shekhawat himself.

He shared with me what he believes helped him survive the numerous successful operations he has been a part of: 'In an operation, planning is everything. If you don't plan well, read and research every aspect of the mission, the chances are you will perish sooner or later. As a leader, you should be extremely prepared before you execute the actual operation. One should avoid being in a rush.'

The terrorists stayed in the occupied village during the daytime and crossed the border at night. They would return to the village at around 4 a.m. The ambush was planned for 3.30 a.m.

The various squads occupied their designated positions, and the 21 Para Special Forces surrounded the village. At around 4.15 a.m., they spotted some of the terrorists walking into the village. The team allowed them to go in. In fact, they waited until more and more terrorists entered the village. This continued until around 6 a.m.

Each second was akin to a lifetime of tension for the team that lay waiting and alert, gripping their weapons tightly, ready to fire in the blink of an eye. When Shekhawat was assured that there were enough terrorists inside the village, he asked his buddy, an excellent sniper, to take a clean shot at the sentry post.

---

39 To provide fire support to the main assault party
40 The team assigned to carry out the assault

He cracked a little joke over the secure-communication set, saying, 'We are going to take a shot at the sentry post first. You all may start after that. People shouldn't say later that 21 Para took hundreds of rounds with them and did not kill a single terrorist—so we should kill at least the sentry for sure.'

But even as Shekhawat's buddy geared up to fire his Dragunov,[41] he spotted another terrorist joining the sentry at the post and doing some stretching exercises. He informed Shekhawat, and the two of them took simultaneous shots at the sentry post, killing both terrorists on the spot.

All hell broke loose! A barrage of bullets was fired from all sides—it was impossible to tell where it was coming from. The Para SF used everything from rocket launchers, general-purpose machine guns called PKMGs, and MGI high-explosive dual-purpose rounds. Many huts caught fire in the crossfire. The terrorists began to flee, though many stayed and fought back. It was utter chaos.

Slowly, the main assault party began moving towards the village under heavy fire, but they were unable to make much progress in the face of non-stop firing. Eventually, Shekhawat asked a troop commander from the 'stop and cut off' party to move behind the terrorists. Taking six boys with him, the troop commander attacked heroically from the rear, as Shekhawat's parties kept the terrorists engaged from the front. From the back, the troop commander managed to break the position of the terrorists in an incredible display of raw courage, coupled with excellent shots from his men. He was later awarded a Kirti Chakra for this unmatchable bravery. Rashtrapati Bhavan had erupted in applause when tales of his valour were narrated.

By the end of the three-hour-long operation, many terrorists had either fled or been neutralized by the 21 Para SF. This particular

---

41 An excellent Russian sniping rifle.

operation made headlines in local newspapers and TV channels, which gushed about the bravery of the commandos for days. The operation and its success strengthened the belief of the locals that the Indian Army was there for them—and that as long as it was around, foreign terrorists would not be allowed to dominate the area.

As he finished telling me this story, Brig Shekhawat took a sip of his tea, leaned back in his leather chair and said, 'Five terrorists were killed, two were injured, and many weapons were eventually recovered. But what gave me greater satisfaction was that there was no loss of life on our side in an operation which was undoubtedly risky. I later recommended my young troop commander's name for the Kirti Chakra and another officer for the Sena Medal. The boys did well.'

What Col S.S. Shekhawat left out, in his characteristic modesty, was that he has also been recipient of the Sena Medal in the same mission. His citation reads,

For consistent exemplary leadership, offensive spirit and personally leading his men from the front, Lieutenant Colonel Saurabh Singh Shekhawat is awarded the Sena Medal (Gallantry).

~

8 September 2008
Operation Loktak Vijay
Loktak Lake
Manipur

Operation Loktak Vijay was a historic operation conducted over water amidst giant shifting grass. The entire operation, under Lt Col S.S. Shekhawat's extraordinary leadership, spanned forty-five days and included continuous surveillance, planning and training in preparation. However, the execution of the plan took only four minutes.

The news of the operation's success spread like wildfire and made Lt Col Shekhawat a national hero and a household name. When Arnab Goswami interviewed him on Times Now in 2011, the TRPs of the channel went through the roof. Even at the time of writing this book in 2021, that interview is among the most-watched YouTube videos, splashed as it was across various social media platforms. I scrolled through the comments on that video and, to me, it seemed almost as though this man, even though he was no Bollywood actor or celebrity, had a cult following among millions of Indian youngsters who swore that they would join the SF and follow his path.

But when I asked him about the interview and the fandom he inspired, Brig Shekhawat seemed shy, and confessed that he does not spend time on YouTube, and even watches the TV news only occasionally. Rather, he invests his free time in building his body or learning a skill. That is his source of entertainment. I wondered about that until I began to research him. That's when I discovered that Brig Shekhawat was capable of being chivalrous and dauntless at once. Being his biographer, I take the liberty of saying that he is a man of contrasts. An avid reader to this day, and an equestrian leagues apart who deeply loves his mare, Corsica, he is poised and chivalrous but can be dangerous and adventurous too. We see these traits of contrasting emotions in his daring operations like Operation Loktak Vijay.

Loktak Lake is one of India's largest freshwater lakes. Historically, Loktak is deemed the fountainhead of Manipuri culture, but there was a time when its pure waters were infested with the disease of terrorism. Various terrorist camps were located at different parts of its banks. The area around the lake was quite densely populated, with around 6,000 locals living in 2,000 huts.[42] Terrorists would extract hefty sums of money from them, killing anyone who refused

---

42 https://www.telegraphindia.com/north-east/loktak-operation-to-continue/cid/637482

to pay up. In sum, Loktak was a forest, a lake and a settlement all rolled into one.

One day in 2008, local policemen informed 21 Para (SF) about a villager who had been abducted by a terrorist group that operated in and around Loktak Lake at the time. The kidnapped man's brother, a local fisherman, had sought help from the police to find him.

Lt Col Shekhawat asked the fisherman to show them the terrorist camp where his brother was likely being held. He complied, but since the lake was vast and the biomass was 10–12 feet thick, there was almost no visibility, and the squad was unable to locate the camp, even though they were using the best binoculars. It was not an option for the Indian Army to sail the lake either, because their presence would have immediately alerted the terrorists. At first glance, the mission seemed like a no-go.

Shekhawat gave a GPS tracker to the fisherman with the intention of gleaning the terrorist camp's coordinates. Since he was a local, no one would suspect his presence among them. Accordingly, the man sailed out in the dark of the night, hiding the GPS in his fishing tokri. He returned to Shekhawat with the coordinates the following morning.

Shekhawat was excited. He pored over the coordinates, plotting them on the map, and took his squad to the nearby Karang Island, which provided them the benefit of a location hidden from terrorist surveillance. The squad trained their binoculars on the coordinates and, within fifteen minutes, they had managed to locate the terrorists: there were around sixteen or seventeen men in uniform, which was more like a shabby version of the Indian Army uniform, armed with AK-47 and M16 rifles.

Now that the enemy had been successfully located, the next hurdle that Shekhawat had to cross was figuring out how to enter the camp. Several options—such as crossing the lake using motorboats and plastic boats—were discussed and discarded. It was reasoned that the boats would get entangled in the fishing nets under the water and the element of surprise would be lost.

Finally, the squad zeroed in on the plan of physically swimming across the lake to reach the camps. There was a canal near the lake. Shekhawat and twelve of his boys began training in the waters of the canal during the night to protect the secrecy of the operation. It was a strenuous plan and, no matter how hard they tried, they could only cover 2–3 kilometres in one go. Many of the boys lost weight during the training period, others vomited, and two of them had to be admitted to the hospital.

Eventually, Shekhawat's men complained to him, saying, 'Saab, even if we manage to swim till the camp, we will be too exhausted to fight. This is a suicidal mission.'

Shekhawat was perturbed, but he had not given up yet. The team finally decided to use the same type of canoe that the locals used on the lake to sail to the terrorist camps.

Though the squad members were trained to fire bullets from the canoe and also to row them, there were still other problems plaguing this risky mission. First, there was the problem of mobility, given that they were surrounded by water and would be unable to quickly make an escape in case of an attack. In addition, they would be exposed while taking the shots, as they had no cover. The men would also not be able to wear bulletproof jackets because their sheer weight would have drowned them if they were to swim. It appeared as if the mission was not only suicidal but also completely against popular military tactics. The odds were stacked against them.

Many of the men backed out in the face of obvious danger. Lt Col Shekhawat said to his boys, 'All of us have our limits. The surprise element is our cover. If we maintain it, we can win—and if we lose it, we will die. If you want to back out, it is okay, and I respect that. But if you want to continue with the mission, it would be done my way.'

Meanwhile, he also met the Corps Commander[43] and briefed him on the plan. His first question was, 'How many civilian huts are there, and are there any chances of civilian casualties?'

Lt Col Shekhawat replied, 'There are four terrorist huts which they are using as their camps. These are located amidst fourteen civilian huts. Since we have already marked the huts, there should be no loss of civilian lives.'

There were also reports of two lady militants, who were the wives of the terrorists in the camps. Since he knew this already, the corps commander asked, 'There are female militants present too—if anything happens to them, Manipur might flare up. I hope you are aware of that?'

Lt Col Shekhawat replied firmly, 'Yes! Things can go wrong, but we will have to take that chance.'

The Corps Commander asked him again about the possible loss of men, and Shekhawat replied, 'Yes, since we are exposed and without any cover, there is a risk.'

The Corps Commander said, 'Shekhawat, your plan is not only bloody audacious but it's also dangerous. Right now, I don't think I can permit you to undertake the operation. Wait for my permission before taking any action, no matter what. This is an order.'

Lt Col Shekhawat saluted him crisply before leaving his office. He was disheartened, but orders were orders. However, another chance would come his way when the same terrorists attacked the battalion headquarters of a Gorkha Rifles. While five Army personnel were injured, no loss was registered on the terrorist side. This could not be tolerated, and the enraged Corps Commander eventually permitted Lt Col Shekhawat to undertake Operation Loktak Vijay.

The elated Shekhawat now awaited the right moment to strike. That time arrived on the morning of 8 September 2008, when the

---

43 An officer ranked at the lieutenent general level, in command of Corps-level troops.

terrorists engaged in a firefight with the Manipur police commando post. There was intense firing from both sides. The surveillance squad (which Shekhawat had earlier established), equipped with LORAS[44] binoculars at Karang Island, now provided him with crucial information about the terrorist movement. By the time it was 8 p.m., they informed Shekhawat about the terrorists' retreat to their camps.

It was now or never.

Brig S.S. Shekhawat paused in the telling of this story to sip on his coffee with gur, which had just arrived along with some fox nuts roasted in ghee and some seasonal fruits. It had been two hours since I had been sipping on green tea, but those two hours had flown past, immersed as I was in the story he was telling me.

Smiling a little at my enthralled expression, he continued, 'So, Admiral[45] Shekhawat's fleet—three canoes carrying twelve men as part of the main assault party—moved out in the moonlit night. Within no time, we were in the vicinity of our target and could see it. However, before launching the attack, I wanted to do one quick perimeter patrol to prevent any collateral damage. I mounted a plank on the canoe, and the canoe flipped. We all went underwater, but since we had trained so well, there was no noise. We adjusted the canoe silently. Our ammunition and rocket launcher were also kept in hyperbaric bags, so they were secured too. Then I felt cold air on my back, which indicated that the air was moving from our direction towards the terrorists', and the chances were that they could have heard the rowing of our oars. So, we turned our canoes and took a detour three kilometres against the wind.'

---

44 A long-range recce and surveillance system which can be used to see long-distance targets even at night.

45 Since the operation was carried out on water using canoes, Brig Shekhawat jokingly refers to himself as the Admiral in this situation. His actual rank at that time was Lieutenant Colonel in the Indian Army.

Finally, when Shekhawat and his men reached the target at around 11.30 p.m., they found a massive chunk of biomass between them and the target. Shekhawat and his buddy dived into the water and climbed the biomass just below the terrorist sentry post. This part of the plan hadn't been discussed beforehand, so they had no clue about what to do in this situation. They waited silently, and when the sentry moved from his position at around 3.30 a.m., Shekhawat immediately radioed his boys to come forward, pull their canoes on the biomass and set the platforms on top canoes to stand and look above the biomass and fire accurately.

By now, dawn was breaking. The surveillance team with their LORAS binoculars had informed Shekhawat about eleven terrorists dispersing. It was go time.

Shekhawat's buddy Lakshman Singh fired the first shot from his flamethrower, an RPO7 rocket launcher meant to destroy concrete bunkers. It pierced through a thatched hut and splashed into the water.

There was silence for a split second before all hell broke loose.

The 21 Para (SF) squad replied with full ferocity. It was a real-life adrenaline-filled action scene that would not have been out of place in a Hollywood movie. But it was all over in a span of four minutes.

As silence fell, the men heard a woman screaming in pain. Entering the hut, they found a female militant, a rifle in her hand and a fragment of hot metal embedded in her calf. Shekhawat asked her to let go of the rifle and promised her that his team would provide medical aid to her. She dropped the weapon.

It was a well-organized camp. They Army team recovered many weapons, including AK-47s, rocket-propelled grenades, thousands of rounds of cartridge, M16 rifles, mortars, and many odd radio sets, as well as mobile phones, solar lights and SIM cards. The then DGP Y. Joykumar also gave a press briefing after the joint operation, talking about how the same terrorists were behind the recent bomb attack at the chief minister's bungalow. It was important to tell people that the operation had broken the backbone of that terrorist group, instilling their faith in the system again.

The 21 Para (SF) had killed nine terrorists that day in a historic operation that had looked almost impossible at one point in time. Shekawat's boys later joked with him, '*Saab, aapne toh humari training mein hi itni le li thi ki main assault toh hume bahut aasaan laga* (You made us train so hard that the main assault felt like a cakewalk).'

Lt Col Shekhawat smiled and patted his men on their backs before telling them, 'The more you sweat in peace, the less you bleed in war. Never forget that.'

Later, the then President of India, Pratibha Devisingh Patil, awarded him with India's second highest peacetime gallantry award, the Kirti Chakra, on 26 January 2009 for the successful completion of the operation.[46]

~

2010
Assam

In retrospect, Brig Saurabh Singh Shekhawat has done well not just on the field, but also in his Army courses, he has also been a part of several UN missions and has commanded his parental unit 21 Para (SF) and continued to lead his men from the front. Under his command, the 21 Para (SF) continued winning laurels for the '*Naam, Namak, Nishan*'.[47] He learnt skydiving and trained to become

---

46 https://pib.gov.in/newsite/PrintRelease.aspx?relid=46977

47 An ethos that calls upon Indian soldiers to strive for the good name of their country, the salt that they have partaken in and the glory of the national flag/regimental standard, to the extent of making the supreme sacrifice of their lives when required.

a combat free faller when he commanded his parental unit. After this period, he was posted to Kashmir.

Lt Gen Satish Dua (retd), former GoC 15 Corps and Chief of Integrated Defence Staff Committee, remembers the Kashmir floods and how Brig Shekhawat would lead rescue operations on his horse. The area was badly waterlogged and there was no way vehicles of any kind could enter. Gen Dua told me, 'He and I were horse-riding partners. But he is a superior horse-rider—he identifies with the horse and can understand everything the majestic animal wants. He rides horses without a saddle as well.'

As I write his story, I can see clearly that this was a man who never settled for the ordinary. He was always in pursuit of something extraordinary, which would benefit his nation and her people. One particular incident left a profound impact on Brig Saurabh Singh Shekhawat and pushed him to eventually achieve something which changed the dynamics of combat skydiving in India forever.

In 2018, Col S.S. Shekhawat, along with his team of paratroopers, including young trainees, was training in Ladakh. During one of the combat free-falls, a soldier who had just joined the forces lost his life after suffering a cardiac arrest right after his first jump. Shekhawat had always been particular about the safety and wellbeing of the men under his command all his life. He realized something was wrong with the training modules. He read the case report of the young soldier's death and concluded that it could have been prevented if the boy had got an opportunity to train themselves in a vertical wind tunnel—something all advanced armies (such as in the UK, the UK and even in Myanmar and Pakistan) do.

He also discovered that the Indian Air Force[48] had been in the process of trying to procure a vertical wind tunnel for the past fifteen

---

48 https://www.tendersontime.com/india/details/vertical-wind-tunnel-trainingairhq2390921-410be69/

years or so, in order to modernize combat free-fall training. However, because of bureaucratic hurdles, this dream had not been achieved till then.

One of the officers, who has worked closely with Brig Shekhawat during the construction of the vertical wind tunnel, revealed to me that 'the vertical wind tunnel was not only Col Shekhawat's passion but also of utmost importance to the nation. I have seen him working around the clock, day and night, to fulfil this dream—which at one point in time looked almost impossible. The construction of the tunnel was not easy but it was revolutionary. Then, he has also trained many tandem masters and created a pool of talent for the country to dip into. This helps us immensely on the battlefield. Imagine taking a doctor, a religious teacher and a sniper (who is not a free faller) into a battlefield without being detected by the enemy system. It can change the whole equation. The vertical wind tunnel speeds up the entire training process along with providing complete safety to the paratroopers. This facility liberates us.'[49]

The officer paused again and took a breath before adding wistfully, 'But do you know, ma'am, that all this had come at a great personal cost? While Col Shekhawat's course mates are commanding their brigades, he held on to his Colonel rank despite being approved for the next rank. Who does that? God knows how many interviews he sought with higher authorities just so that he could avoid a promotion because he feared his posting would delay the construction of an important project for the nation. I am glad I served under him and learnt what it means to live up to the Chetwode Motto.'

---

49 These techniques were not taught in India. He had visited many foreign countries on his own expense to learn various techniques of sky diving. Once back, he taught those to the men under his command, who further train more paratroopers. Thus, creating a pool of talent.

These new concepts of tandem diving and vertical-wind-tunnel training have negated the need for outdated combat free-fall training techniques. This enables our country to train the finest, most advanced paratroopers, combat free fallers, SF operatives, National Security Guard commandos and Air Force officers, who would prove to be game changers on the battlefield in the days to come, not to mention the many families who would be saved from dealing with the loss of their loved ones.[50]

When I asked Brig S.S. Shekhawat about this, he gave the entire credit to his senior officers, Maj Gen G.S. Bisht, Maj Gen Sanjay Singh, Col Vikas Singh and then also to the National Security Advisor (NSA) Ajit Doval, KC, and ultimately Prime Minister Narendra Modi.

Brig Shekhawat told me, 'The tunnel was pending for a long time, but the stars were aligned, and I was fortunate to have supportive seniors who pushed hard for it. I was only the ground soldier. I want to mention the prime minister's generosity when the higher authorities gave him a presentation on the tunnel. He asked only one question: '*Isse kya hoga* (How will this help?)?' He cleared the project immediately when they told him that it would help in saving the lives of several soldiers each year. This is how our first vertical wind tunnel came to be.'

NSA Ajit Doval, KC, inaugurated the facility in July 2021 and proudly dedicated it to the military. It is part of the prime minister's vision to modernize the Indian military to make India a "super military power," thereby making India a superpower. Today,

---

50 The Government of India is now taking aerosport seriously. It has enacted many skydiving friendly laws, which will hopefully change the aerosport scenario in India and establish aerosport-related infrastructural facility in the country. This will also enable even civilians to learn this extreme sport in India itself.

paratroopers from across the country come and train in the tunnel. By the time this book is published, the first vertical wind tunnel for the public would also have been constructed in Hyderabad by GravityZip, becoming India's first indoor skydiving arena. The Indian Air Force also came up with an advanced Hypersonic Wind Tunnel (HWT) test facility in 2022.[51]

Still, the vision that Brig Shekhawat has had for the next generation of paratroopers, in the face of so many hurdles, has been nothing short of remarkable. And the tunnel has the singular honour of being the first vertical wind tunnel in India. I made a mental note on how his passion for his job had pushed him to achieve many firsts, from Mount Everest expeditions and techniques of skydiving to a whole infrastructural facility. This man is a superstar. And like most superstars, his life is not untouched by controversy.

I asked his wife, Renuka Rathore, about the controversy he had faced in the past.[52] She laughed it off, saying, 'When he was dragged into an unnecessary controversy where he had no role to play, he was hurt, but I asked him, "Why are you so upset? What is the maximum that can happen? That you will retire as a colonel? That's it." He told me he was not bitter about his rank. It truly does not matter to him. He said, "I am a soldier, I only want active soldiering. And I need

---

51 https://economictimes.indiatimes.com/news/defence/rajnath-singh-inaugurates-indias-first-hypersonic-wind-tunnel-test-facility/articleshow/79817803.cms?from=mdr

52 A controversy finds a mention on Wikipedia. Interestingly, the Wikipedia page on him is filled with false information, with errors in his mother's name, military career, type of medals, his date of birth, commands he held and more. The pages have quoted from YouTube and newspaper articles, on which thousands of other YouTube videos have been based. None of them ever bothered to approach the Indian Army for clarification but simply fake news one after the other. He wrote several times to Wikipedia to remove the false information but they rejected his request for reasons known only to them.

peace to focus on my job. People's unnecessary speculations trouble me more.'"

I also had a word with Lt Gen A.K. Singh (retd), ex-Southern Commander and the former Lieutenant Governor of Andaman and Nicobar Islands and Puducherry. He has known Brig Saurabh Singh Shekhawat for many years. He told me, 'Saurabh is a man with true soldierly attributes. His men would follow him anywhere only because he puts himself in harm's way first. Anyway, I would want my battlefield subordinates to be like him. He exhibits excellent abilities and agility during any task. He is daring and courageous, but at the same time, he is responsible and calm. This shows how he has found his balance. Tell me, what kind of person would climb Mount Everest thrice! It is only a man with the utmost determination and skill and also a man who is not reckless who would do this. If you see, in Loktak, he trained for forty-five days along with his men—that kind of patience is rare, I tell you. If a person with his credentials does not make it to the next rank, then who would? The Army has a very fair and transparent system of rectifying itself. It has a robust redressal mechanism which duly addresses various grievances. So, you can see what happened eventually—he was promoted, he got his dues.'

Telling Brig Shekhawat's story has not been easy. His story has several layers; stories of immense valour and adventure and dangerous odds. It took me several months just to learn about the areas of mountaineering, skydiving and others he had operated in. It made me realize that his life is the best portrayal of the opportunities and adventure the Indian SF offer as an incredible career option. Brig Saurabh Singh Shekhawat, KC, SC, SM, VSM, is an icon who inspires a whole nation. I hope with this story, I will be successful in giving this nation the Indian version of Captain America or Napoleon Bonaparte. Why follow their ideals when India is a land of homegrown superheroes? I also hope more youngsters venture into the field of extreme sports and test their limits, and that cadets training in military academies find motivation in his story, as do defence aspirants. As storytellers, it is our duty to bring forward such inspiring tales from the soil of India to give our children the right role models to seek hope and courage, like we do from the tales of Netaji Subhas Chandra Bose or Chandrashekar Azad.

~

This story is based on interviews conducted with Brig Saurabh Singh Shekhawat, who is now commanding a brigade, and with the people in his life.

His family—his wife, Dr Renuka Rathore; his mother, Dr Shraddha Chauhan; his father, Dr Jaswant Singh Shekhawat; and his daughters—were generous enough to share their stories with me.

I also spoke at length with the colleagues who served under Brig Shekhawat, his senior officers who cannot be named (for obvious reasons), as well as with Lt Gen Satish Dua (now retd) and Lt Gen A.K. Singh (now retd), who all provided valuable inputs. Maj Gaurav Arya and other course mates who are still in the service also helped me immensely.

Brig Saurabh Singh Shekhawat's is a fascinating story that took me several months to write—it is a story which demanded sweat and toil, and proved to be the most challenging in my entire writing career so far. His legacy glorifies the rare feats SF officers perform in the filed of extreme sports and other challenging tasks, along with performing their regular duties. Through his story, I felt the need to tell people of India such SF operatives are bigger than any sports or movie star, and we must change our idols to achieve more. It is also by the grace of God that he has fortunately survived, despite dangerous odds, to share his tales of raw courage.

No book on the Special Forces can be complete without mentioning Brig Saurabh Singh Shekhawat's name. He is our own superhero.

9 Para Special Forces: The Ghost Operators of the Valley

THE 9 PARA Special Forces is popularly known as the Ghost Operators of the Valley. Mountain Rats, Pirates, Headhunters'[1] are other popular names. They are everywhere, but they are invisible. As a unit, their reputation precedes them, with tales of hair-raising missions and incredible bravery.

With four Ashoka Chakras, one Kirti Chakra, seven Vir Chakras, twenty-four Shaurya Chakras, 111 Sena Medals, five Vishisht Seva Medals, six Yudh Seva Medals, fifteen Mentions-in-Dispatches, 116 Chief of the Army Staff commendation cards,[2] and many other medals, awards and honours—enough to equip a medal mint—their achievements put the unit at a stature difficult to reach, let alone compete with.

---

1   The Naga regiment is usually related with the Headhunters name, but there has been stories of 9 Para chopping off the heads of militants. During interviews, many outsiders referred to them as headhunters.

2   From data available till 2016.

The 9 Para Commando Force was raised on 1 July 1966 in Gwalior. They were designated as the first Commando in the Indian Army on 15 January 1969 and awarded the Balidan badge on 10 May 1972. The word 'Commando' was changed to 'Special Forces' in 1994. They have also been the recipients of several battle honours and theatre honours for the 1971 and Kargil wars, and also the 'Bravest of the Brave' Award on 26 January 2001.

From Operations Rakshak, Parakram and Vijay, expeditions to treacherous mountains, earning medals in international sporting events to protecting the Prime Minister of India in 1985 as the first inner-security force; 9 Para (SF) have recorded a wide range of achievements in the history of the SF.

The unit has continuously been in active operational command, at and even across the Line of Control and the Line of Actual Control, making them an ever-ready force capable of inflicting severe damage to enemy forces and morale. There are many accounts of legends that send a chill down the spine. For instance, in the 1990s, when a band of Nines cut off the heads of terrorists and threw them at the feet of their livid army commander. Their training standards, which they claim are higher than most SF units because of operations undertaken in high snowy mountains, require great agility, mental strength and fitness. They have kept their attrition close to 99.9 per cent.

Indian children have grown up hearing many legendary names from 9 Para (SF), such as Maj Sudhir Walia; Captain Arun Singh Jasrotia; and Paratrooper Sanjog Chhetri. The officers' mess of the battalion is studded with pictures of these heroes—with new generations of officers sharing drinks and laughter under their watchful gaze.

The 9 Para (SF) is a league of extraordinary soldiers where every man is a hero, and stories abound of unimaginable bravery in the most difficult of times. I was privileged enough to write about the modern military heroes from Nine—be it Subedar Major/Honorary

Capt Mahendra Singh, KC, SM, whose valour will make anyone gasp; Capt Tushar Mahajan, SC (posthumous), the young daredevil operative who was an enigma in every sense; or Maj Manish Singh, SC, whose never-say-die attitude will give you the hope to fight your miseries.

I would also like to mention the exemplary cohesion and warmth that characterizes this unit. The officers of 9 Para (SF) are greatly concerned for their men and their families. Their exclusive, one-of-a-kind 'Killed in Action Cell', where a religious teacher extends all kinds of moral support, in addition to other efforts they undertake for the families of their lost comrades, also touched my heart deeply. The ladies behind these men are simple and possess rock-solid strength.

These extraordinary stories from one of the world's most battle-hardened, combat-rich units were written following back-breaking research. I was privileged enough to have seen the ghosts with my own eyes. I hope I did justice to their stature, and I also hope you realize you are getting a rare peek into their lives.

Jai Hind!

3

Sub Major/Honorary Captain Mahendra Singh, KC, SM: The Great Mountain of 9 Para (Special Forces)

2 September 2015
Haphruda forest
Jammu and Kashmir

THE NIGHT WAS darker than a black panther. There were no stars. The deadly jungles of the Haphruda region were overgrown by pine trees, enormous shrubs and bushes. It was an arduous terrain which inspired dread and fear in those who sought to take it on.

However, on that night, there was an air of tension in the jungle. The men from 9 Para (SF) were on the prowl, laying an ambush deep inside the forest. Six squads consisting of thirty-six commandos were taking their positions at and around the inverted V-shaped hill in the middle of the jungle, surrounded by thick, dense trees, giant boulders and small irksome ridgelines—a spot that could not even be traced on a map.

Subedar Mahendra Singh was installing lethal Claymore mines in tandem, perfecting the layout of the ambush laid by 9 Para (SF). Four Claymore mines were interlinked through the wires and were supposed to be detonated together through a single initiation trigger. With the Team Commander,[1] he had discussed how the mines would be detonated as soon as the terrorists entered the ambush area, only to be followed by a spray of bullets that would ensure a complete kill without any casualties on the forces' side.

Mahendra saab's experience and exposure to counter-insurgency operations were of great value to the unit. This senior-ranking Junior Commissioned Officer (JCO) of 9 Para (SF) was an asset in the true sense. He had joined the unit at the peak of militancy in Kashmir Valley in the 1990s, surviving every operation he was a part of with his sheer intellect, planning, courage and, most importantly, his calmness. Yes, he was most definitely a towering figure on whom the team commanders and commanding officers of 9 Para (SF) thoroughly leaned on during the most challenging operations.

Since Mahendra saab had been promoted to the rank of senior JCO in the unit in 2012, he had not been part of active field operations. Planning and managing logistics to support the field troops took up most of his time.

For several days, 9 Para (SF) was involved in many back-to-back operations. Saab[2] thought about accompanying the team that was scheduled to leave for operations in the Haphruda forest, which had a history of heavy casualties because of how dense it was, with its difficult ridgelines and the zero visibility it offered. Forces operating

---

1 The leader of the entire team, with several Troops Commanders under him, responsible for smaller squads comprising JCOs and ORs.
2 What JCOs in the Indian Army are respectfully called

in buddy pairs would usually find it difficult and dangerous to operate in such an area. Whether in light of the veergati[3] of legendary 9 Para (SF) soldiers like Capt Jasrotia or Maj Sudhir Walia, or other casualties from units which included Maj Mohit Sharma from 1 Para (SF), the Haphruda forest was seen as a jinxed spot for the forces.

The area lies across three major infiltration routes, converging into the same forest. Any foreign terrorist crossing the Line of Control (LoC) to enter Indian soil at the Handwara general area has to traverse the jungle. The Indian Army keeps a strict vigil in this particular forest and the routes around it.

That day in 2015, there was accurate information received about four divisional commanders of a terrorist group getting ready to enter the forest. They were reportedly carrying heavy cash and modern weapons meant to be distributed to various smaller terrorists and over-ground workers (OGWs)[4] who worked for the tanjeem.[5]

The four terrorist divisional commanders were also supposed to be planning something explosive after entering and then scattering across Kashmir to their predetermined areas of operations. These commanders were dangerous and were considered to be huge assets of the organization to which they belonged. In that capacity, they were also supposed to radicalize and train local Kashmiri youth for more extensive operations in the future.

The top brass of the terrorist group in Pakistan were sure of the four commanders' operational success, but they did not know that the ghosts—the 9 Para (SF)—were keeping close vigil over the tanjeem. The unit, reputed for the sheer brilliance of its human

---

3   When one sacrifices his life in the line of duty in the hands of enemy for the nation
4   Small-time helpers or informants hired by terrorist organizations, who run various errands in the particular area
5   Terrorist organizations. Here, it refers to the Lashkar-e-Taiba.

information network, technological edge and professionalism, had not only intercepted various conversations between the top terrorist operatives and the local commanders regarding the infiltration of foreign terrorists, but it had also sent in two of their operatives posing as local OGWs or guides to infiltrate the whole chain of command.

The task for infiltration of foreign terrorists was assigned by the local area commander to the OGWs working in the area, who was blissfully unaware of the actual identity of the SF operatives working for him. This ensured that the 9 Para (SF) kept receiving accurate information about the four Pakistani terrorists.

In the Ops room of 9 Para (SF), in the presence of Mahendra saab, a brilliant plan was hatched to encounter the terrorists in the Haphruda forest itself.

Sometimes, however, the most brilliantly laid out plans don't work out, indicating there are larger universal forces at play that exercise complete control over a person's destiny. This is what happened during the Operation, when 9 Para (SF) lost Lance Naik Mohan Nath Goswami and had to bear the brunt of the legendary Mahendra saab getting injured. Many others from 9 Para (SF) also suffered splinter injuries.

The Team Commander remembers how saab had casually mentioned the curse of Haphruda forest and casualties in the past: 'After we decided to carry out a ninety-six-hour-long surveillance-cum-ambush plan, Saab came to me and told me lightly how 9 Para (SF) had suffered legendary casualties in the same forest. But I dismissed it saying we had better weapons and technology now.'

After a pause, the Team Commander continued, 'I had been his Troop Commander[6] when I joined the forces. Our association went back a long way, and his volunteering for the Operation provided

---

6   The junior officer with squads under him comprising JCOs and ORs. There can be several Troop Commanders under a Team Commander.

us with the opportunity to relive the golden days of action we saw together. It was truly unfortunate that Saab was hurt on my watch. As a leader of the pack, any casualty or permanent injury on your men leave a mark on your soul and often don't allow you to sleep at night. After all, we are talking about Mahendra Saab, the great mountain of 9 Para (SF). We have all trained under him, and he got shot during my operation. I have no words to describe how I feel about it. Losing Goswami was equally painful. He was like a brother to me.'

After the ambush was laid and the SF operatives had occupied their positions, the eager wait for the terrorists began. The men were prepared, and there were confirmed inputs—but contact was only a matter of chance. At around 8.30 p.m., they heard some footsteps and observed some movement in the area near the depression of the inverted V-shaped hill on which they had laid the ambush. It was precisely then that Subedar Mahendra saab grinned smugly to himself. The terrorists were marching straight towards the Claymore mines, and their end was just a few steps away. Or so he thought.

As it was, luck ended up favouring the terrorists. Only one of the mines detonated. The rest diffused, registering some technical errors. The divisional commanders were battle-hardened. One of the terrorists was severely injured by the splinters, but eyewitnesses confirmed that none of them heard a single cry or even a whimper from them. As soon as the Claymore mine detonated, an intense firefight ensued between the terrorists and the forces. By now, the terrorists had taken cover and were hidden by the thick cover of the pine trees. The SF operatives in the area were unable to get a clear shot. Still, grenades were lobbed from both sides. A spray of bullets found their way into the action as well. After an hour, both sides stopped firing and decided to wait and watch.

An uneasy silence engulfed the air.

There was not even the slightest sound from the terrorists' side—the 9 Para (SF) operatives thought that the terrorists were probably dead. But the battle-weary Subedar Mahendra Singh and the Team Commander decided to wait till the first light of morning for their search operation to recover the dead bodies and ordered the squads to retain their positions. It was standard operating procedure.

However, around 11 p.m., Mahendra saab heard a beeping sound just below where his squad was positioned. He immediately radioed his Team Commander, and they discussed what this sound could possibly be. There was a chance that the sound was coming from a preset alarm on one of the dead terrorists' watches. There was also an equal chance that some of the terrorists could still be alive. These were, after all, no ordinary terrorists. They were well-trained and highly capable of operating in the same way the SF do.

It was an alarming situation. It was possible that while the forces were waiting for the morning light to hit the jungle so they could properly carry out their search operations, the terrorists may also have been waiting patiently for a clear shot or, even more alarmingly, they might have escaped into the darkness of the jungle.

The Indian Army works on the principle of minimum casualties, where each man's life is precious—but the terrorists who cross the LoC come with the expectation of dying. Causing maximum damage at any cost is their prime focus. Both situations were unacceptable for the forces.

Lance Naik Mohan Nath Goswami was lying with his squad just above Subedar Mahendra saab's squad on the hill. Goswami was a tall, muscular boy from Nainital who had killed eleven terrorists in as many days, over the course of three successful counter-insurgency operations in Kashmir that happened just before this particular Operation.

One of the soldiers from the squad positioned above[7] Lance Naik Goswami's position activated night sight[8] on his Tavor TAR-21, an Israeli bullpup assault rifle, to detect the presence of terrorists. Unfortunately, what this soldier did not know was that the terrorists he was looking for also had night sights and could in fact see through the darkness much more than him.

Later, when the bodies of the terrorists were recovered, they were identified as being divisional commanders in possession of high-quality American weapons, captured from American troops in Afghanistan and eventually passed on to Pakistani terrorists via their Afghan friends.

Meanwhile, as the terrorists spied on the soldier who was watching them, they sent a spray of bullets in his direction. Some of the bullets hit the soldier, injuring him badly. The squad was now clearly a visible target for the terrorists. A grenade fell into the middle of Lance Naik Mohan Nath Goswami's squad. Before it could detonate, Goswami kicked the grenade away, but not before the grenade went off, injuring him and his buddy.

It had become a close-range jungle fight.

Bullets were fired from both sides. Grenades were lobbed without mercy. At one point, Subedar Mahendra Singh radioed the team commander, 'Saab, multiple splinter injuries have been sustained by two of our chaps. I am going to evacuate them. Over.'

---

7   The ambush was laid to cover both sides of the inverted V-shaped ridgeline in the form of small squads occupying heights, one above the other, after a certain distance. Mahendra saab was below Lance Naik Goswami's squad. The first shots hit the soldiers above Goswami's squad.

8   Replacement sights used at night that provide complete freedom to see in the dark with bright infrared visuals

The Team Commander replied, 'No! Saab! That position is exposed. There is a high threat. Someone else will go, or we will wait. It is risky.'

Mahendra saab said, 'Saab, they have radioed me and are writhing in pain. I have promised to bring them down safely. I can see their position. I am nearest to them. Goswami's squad is engaging with the terrorists. Let me bring them down to where I am and then you can send some chaps to evacuate them.'

Eventually, the Team Commander agreed. Immediately, the brave forty-one-year-old Mahendra saab crawled upwards to an exposed position, where bullets were hitting every part of the trees. The injured soldiers were hiding under a thin cover when Mahendra saab reached near them.

Unfortunately, the terrorists' night sights had picked up this new attempt at evacuation too. Firing resumed. Mahendra saab returned fire in a display of raw aggression, managing to eliminate one terrorist at close quarters. This provided him with a narrow window of opportunity, and he immediately managed to successfully evacuate the injured soldiers to where Lance Naik Goswami's squad was hiding down the slope.

However, they could still be seen. Two other terrorists began firing indiscriminately with their powerful 40 mm under-barrel grenade launchers.

Lance Naik Goswami was injured but kept firing back at the terrorists. Only when one of the bullets tore through his waist and another went clean through his heart did he fall to the ground, his hands still wrapped around his gun. He had displayed an incredible display of valour for which he would be awarded the Ashoka Chakra posthumously, India's highest peacetime gallantry award for most conspicuous bravery or some daring or pre-eminent valour or self-sacrifice other than in the face of the enemy

Shocked that Goswami was hit, Mahendra saab looked in his direction for a split second and was immediately hit by a bullet too, which went through his abdomen and straight to his spine. Immediately paralysed, he collapsed to the ground as well. He must have been in enormous pain. Still, his comrades told me that Mahendra saab barely flinched.

The terrorists were now closing in, and despite the fact that Mahendra saab was grievously wounded and on the ground, he kept firing. He shot one terrorist at point-blank range and tried to struggle to his feet. But the pain was too intense. With great difficulty, he turned his head and found his injured comrades still breathing. As long as they remained undercover, he thought hazily, they would be safe.

But the Subedar's task was not over yet. He needed to inform the Team Commander of the developments, so he radioed him again, saying, 'Saab, two terrorists are confirmed eliminated. It was close combat, but we have also sustained casualties. Goswami is no more, and I have also been hit in the abdomen. But the evacuation was successful.'

This was devastating news for the Team Commander. Losing any comrade on the battlefield is something all leaders loathe—and losing one of the finest is always catastrophic. He replied, 'Saab, I am immediately sending some chaps to evacuate you and the others. Please hold on tight. Do not give up, I repeat, do not give up. Nothing will happen to you.'

Mahendra saab replied in a frail voice, 'No, Saab! I forbid it. The position of the other two terrorists is still unknown. They might have been waiting for an opportunity like this. Our positions are exposed, and I think they are also using night sights. Sending more troops will only mean more casualties. I will come down myself.'

The extent of Mahendra saab's injury was unknown to the Team Commander. He did not know the bullet had hit the JCO's spine. He insisted again that he would send for help, but Mahendra saab again refused to accept any attempt to evacuate him.

With near-impossible courage, he crawled forward using his hands. Eventually, he rolled down safely, but finding a safe place to hide in the Haphruda forest, in the middle of a full-blown gunfight, was no easy feat. Still, news of the situation had been passed on, and more squads were closing in. The other two men had also been successfully evacuated by then. The body of Lance Naik Mohan Nath Goswami had also been brought down carefully. The Team Commander rushed to Subedar Mahendra Singh, his pillar of strength. It was unbearable to see him in such pain and so grievously injured.

The Team Commander reassured him, saying, 'Saab! Saab! Don't worry! Everyone has been safely evacuated. In all probability, all the terrorists have been killed, but we will search only at first light. The situation is under control. Please don't lose hope.'

When Mahendra saab realized that the situation was under control and everyone in his team was safe, something inside him changed. The warrior vanished, his job done. Finally allowing the pain and agony to sneak in, he cried out in pain. It the first time he had made any kind of sound since he had been hit. The only thing he managed to say as he was lifted into the nearest bulletproof vehicle, bound for the hospital, was, 'Saab, the pain is unbearable. Please shoot me.'

The vehicle's doors slammed shut at that moment, and so did his eyes. Tears rolled down his cheeks, thinking of all that had passed, the moments long past. A man's entire life flashes before his eyes when he knows he is about to die. Mahendra saab drifted back in time—to the beautiful golden desert and the camels of his homeland.

~

15 January 1974
Mahendra Saab's village
Jhunjhunu
Rajasthan

Mohanlal and Patashi Devi had been blessed with a healthy baby boy, and their happiness knew no bounds. Their second son had been born after two girls, and his father had distributed sweets to the entire village. Everyone congratulated Mohanlal for another son who would no doubt help the family in the harsh desert fields.

Sandstorms and sand dunes were common in their small hamlet, which comprised forty houses. Each household worked tirelessly from dawn till dusk to put food on the table. Extra hands were always welcome.

Of course, nobody knew then that Mohanlal's second son, named Mahendra after the great Lord Indra, the king of all gods, would bring unimaginable laurels and prestige to not only the village but to the entire district of Jhunjhunu as well with his fearless spirit and unimaginable bravery. Nobody had a clue that one day the entire village would gather at Mohanlal's house to catch a glimpse of Subedar Major/Honorary Capt Mahendra Singh, KC, SM. They did not know that the then President of India, Sri Pranab Mukherjee himself, would lean down to award Mahendra's wheelchair-bound frame and salute the undaunted spirit of the man who is a recipient of India's second highest peacetime gallantry award the Kirti Chakra, the only one that 9 Para (SF) has to date. They did not know then that he would be one of the brightest stars of the legendary SF unit.

From the beginning, Mahendra was a calm child who started ploughing the fields with the help of camels at the age of six. He dealt with the creatures with ease and confidence even at that tender age. He had a deep respect for nature and animals. He was studious, diligent and hard-working. His mother was a strong and determined woman who deeply impacted young Mahendra's value system. She would not only look after their house, his five siblings and their paralysed father, but she also worked in the fields.

Subedar Major Mahendra Singh told me fondly, 'Madam, those were simpler times. We were not wealthy, but we were content. I remember our mother feeding us bajre ki roti with a generous dollop of home-made ghee, gur, pickles, and chaach early in the morning. Along with my brothers, I would devour those rotis before we left for the fields. Those rotis, cooked on the chulha, which we ate with relish and gratitude, were priceless. We worked in the fields and also attended the village school located two kilometres away. We covered that distance daily on foot. We never indulged in any mischief, unlike the other village boys, simply because we had no time, and survival was our bigger quest. The entire village would praise my mother for how she had raised such obedient sons. That filled her heart with pride. Her veil covered half her face while she nodded and smiled, still engrossed in her work. She was beautiful.'

Today, they are considered one of the wealthiest families in the neighbourhood because of the efforts of all the brothers. All of them testify to the simplicity and beauty of the older times, when they had nothing except each other. All the kids would sleep together in their modest two-room kaccha makaan, studying by the light of a lantern by night as there was no electricity in the village.

Their father had passed away at an early age, and Subedar Mahendra's elder brother, also the eldest in the family, joined 9 Para Commando force[9] in the 1980s, serving until 1995. He was twelve years older than Mahendra and was, as a result, almost a father figure to him. Young Mahendra always looked up to his brother with deep awe. His maroon beret and badges always brought a sparkle to Mahendra's eyes. When Havildar Ram Niwas was awarded a Mention-in-Despatches for his gallantry during Operation Pawan,

---

9 Until 1994, SF battalions were known as Commando battalions.

his admiration for his brother grew by leaps and bounds. Mahendra who, by then, had grown into a tall, muscular and robust young eighteen-year-old, had recently topped his class. It was then that he decided to join the Commando force.

The family's financial condition was not good, so it was a matter of relief when Mahendra expressed his willingness to join the forces instead of continuing with his studies.

~

October 1990
The Parachute Regiment Training Centre (PRTC)
Agra
Uttar Pradesh

Havildar Ramvilas had hurriedly dropped his brother Mahendra to his friend's house. The friend was posted in a Para unit in Agra for a recruitment rally at the PRTC.[10]

Ramvilas was posted in Jammu and Kashmir and could not manage long leaves to stay with his younger brother Mahendra. But the eighteen-year-old Mahendra, who had just left his home for the first time in his life, was mesmerized. He stayed with the Para soldiers, who impressed him with their energy, focus and determination. He also rose at 4 a.m. with the other soldiers and ran long distances with them. At the outset, it was not a new lifestyle for him. After all, a life in Jhunjhunu had inured him to hard work. He loved to run—it made him feel free and gave him great joy.

The recruitment rally was just around the corner, and village boys like him did not get second chances. He knew his brother

---

10 Training at PRTC is the first step for members of other ranks aspiring to become paratroopers. The centre is now in Bengaluru.

had spent a good amount of money on him, and he could not disappoint him.

Mahendra also wanted to buy beautiful clothes for his mother and sisters, whom he had seen in the same old garments year after year. Mahendra was aware that his full-time employment would make a big difference to their quality of life, reducing the burden on his adored elder brother.

The month flew by like the wind. Finally, D-Day arrived. It was the beginning of autumn in October—the leaves looked like flowers, the earth was ripe and the air was cool. Yet, Mahendra noticed nothing. He was a bundle of nerves that day. A crowd of thousands of young men from Uttar Pradesh, Haryana and Rajasthan had gathered on the open grounds. They were divided into groups of 100–150 and made to run. Mahendra knew that he had a natural talent for running, but the problem was that there were too few slots for the many thousands who had showed up that day. He prayed to his gods and thought of his mother desperately.

He must have underestimated himself that day, because he shared with me, 'Madam, I came first. Since I was always good in academics, I had also cleared my written exam. And I cleared my medical smoothly too. With each milestone, my confidence level rose. My overall cut-offs were also higher than those of the other boys. Eventually, I was selected. That day was perhaps the happiest day of my life. I still remember the date—it was 21 October 1990. I was recruited and had to start my six months of basic training at the Parachute Regiment training Centre, though I knew that the maroon beret was still far away. But I had to wear that just like my brother.'

After the basic training at PRTC in Agra, Mahendra was supposed to start specialized training as a probationer at a Para Commando unit. That was his gateway to the maroon beret. His day, along with those of the other recruits, started at unearthly hours with rigorous physical training and drills. Mahendra also learned to assemble and disassemble various guns and weapons. He was also taught various forms of unarmed combat and offensive attacks.

The ustads were happy. The 'josh' is the foundation of a soldier, it makes him who he is—and young Mahendra displayed plenty of it. No wonder he cleared his basic training with flying colours. But the young probationer would hardly get time to sleep. Four straight hours of sleep on Sundays was a huge luxury for them. But Mahendra danced to the tune of his fiery heart that pushed him not just to perform but to excel.

The advanced training was more rigid and rigorous, with sleep time even further reduced. There was battle-obstacle training, where Mahendra would climb ropes, jump on cars, crawl in muddy pits and jump through rings of fire. The recruits were trained in unarmed combat skills and taught all about raids, ambushes, firing and the usage of hand grenades. Mahendra realized these things excited him. The eagerness to take risks and seek adventures came naturally to him. His fellow recruits often struggled, but he excelled in every aspect.

The day eventually came when he swore an oath on the Bhagavad Gita to serve the nation during his Kasam Parade. It was a glorious day. The young recruits who had finally made it were decked out in crisp uniforms and shiny boots. Their heads were held high as they swore an oath on the Gita, the Quran, the Bible and other holy books, depending on their faith, during their Kasam Parade.[11] They were now trained soldiers. Among them was young Mahendra, who was excitedly gearing up for probation with the 9 Para Commando Forces.

---

11 The Kasam Parade is held for the recruits who take an oath on their religious books to serve the nation and to lay down his life, if need be, in upholding the nation's honour. The soldier is further posted to a battalion or regiment for training; this is equivalent to the passing-out parade of officer-cadets.

Though he had no illusions, Mahendra knew his biggest challenge was just about to unfold in the form of the gruelling ninety-day-long 9 Para probation course. His abilities would be tested to the core.

~

September 1991
Jammu and Kashmir

The combat-proven and dependable Mi-8 helicopter could be seen above the canopy of the gigantic trees in the dense jungle. The noise of the rotors of the Mi-8 echoed in the wilderness. The young soldiers were ready to take their first slithering[12] lessons. Mahendra was wearing his combat uniform and carrying his primary and secondary weapons along with his combat kit, which weighed around 20 kilograms. He adjusted his helmet one more time, checked his slithering gloves and then slithered effortlessly using the rope attached to the chopper. Not once did he hesitate. Instead, he seemed to visibly enjoy the process, landing precisely at his designated spot. Glancing upwards, Mahendra smirked a bit, remembering the strenuous probation routine he had been following for the past few days. The moment he entered the coveted premises of 9 Para Commando force, he had become one with the hundred other probationers who had also volunteered to join the commando force. However, within two months, they were only thirty-three of them left.

The training schedule was designed to break every probationer into pieces, such that they could never imagine earning a Balidan badge. Mahendra had pledged to himself that he would either earn the badge and become a part of the legendary unit, or his dead body would be wheeled out of the battalion's main gates.

---

12 An exercise where troops are dropped from a helicopter in an area of
   operation using a rope attached to the chopper

There were times during speed marches when his feet would bleed, moments during mountain climbing when his senses would go numb and his body would ache as he crawled over the rough terrain. Still, he refused to give up. He knew that acquiring the Balidan badge demanded a certain toughness, an ability to conquer mental blocks, a lot of patience and extreme stamina to prepare soldiers for unconventional warfare. The badge tells them that you, as a man, no longer exist—only the soldier remains, for whom the sacrifice of the self is the ultimate dharma in the nation's service.

In the end, by December 1991, out of the original hundred volunteers, only twenty-five had been selected for the 9 Para Commando Force. They would receive their maroon berets in a marooning ceremony. Mahendra was now finally one of them—he was Paratrooper Mahendra Singh.

Mahendra met his elder brother Havildar Ramvilas for the first time since his training began. They looked at each other with newfound admiration. Mahendra knew his elder brother was in a position where he himself aspired to reach one day. On his part, Havildar Ramvilas knew now that Mahendra had been marooned, his younger brother would undoubtedly carry forward his legacy of successful operations.

~

2021

Subedar Maj/Honorary Capt Mahendra Singh, KC, SM, is one of the few soldiers of the 9 Para (SF) who has been with the unit since it was first deployed in the Kashmir Valley during the exodus of the Kashmiri Pandits to today's times, when the Indian Army and other Indian security forces have reasonably controlled militancy in the region.

Parents of Brigadier S.S. Shekhawat, Dr Shraddha Chauhan and Dr Jaswant Singh Shekhawat, with the author in 2022

Brigadier S.S. Shekhawat, when he was a colonel ranked officer, with the author in 2021

Brigadier S.S. Shekhawat with his wife and daughters

The author before the roll of honour board that proudly carries the photographs of the operatives who had made the supreme sacrifice in Operation Rangdori Baihk

Major Manish Singh, during his action days, with his band of brothers from 9 Para (SF) in the Kashmir Valley

Captain Tushar Mahajan, a.k.a Tamir, as a covert operative

A fully equipped Captain Tushar Mahajan, before leaving for a difficult operation with his troop

Subedar Sanjiv Kumar, Havildar Devendra Singh, Ptr Balkrishan, Ptr Amit Kumar
and Ptr Chattrapal Singh, who made supreme sacrifice in Operation Rangdori
Baihk in 2020

Colonel Santosh
Yashwant Mahadik as a
Commanding Officer

Colonel Santosh Mahadik sharing a meal with his buddy
during one of their ambushes in a deep jungle

The Jammu & Kashmir Entrepreneurship Development Institute building during the Pampore Operation in 2016

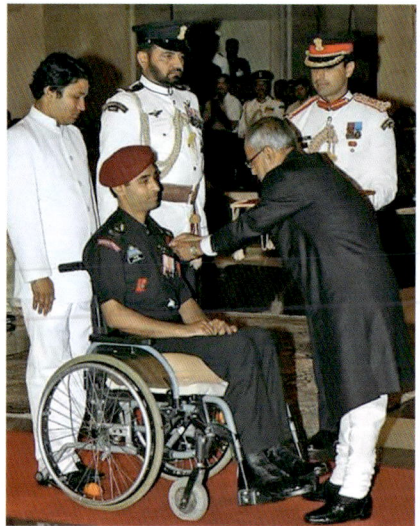

Lieutenant Swati Mahadik, wife of Colonel Santosh Mahadik, with her children Kartikee and Swaraj at her Passing Out Parade

Major Manish Singh being awarded the Shaurya Chakra, third highest peacetime gallantry award in India

Subedar Major/ Honorary Captain Mahendra Singh being awarded the Kirti Chakra, the second highest peacetime gallantry award in India

Major S.S. Shekhawat being awarded the Shaurya Chakra

Subedar Major/ Honorary Captain Mahendra Singh during his action days at a forward post near the LOC

Subedar Major/ Honorary Captain Mahendra Singh with his wife Savitri and children Pradeep and Suman

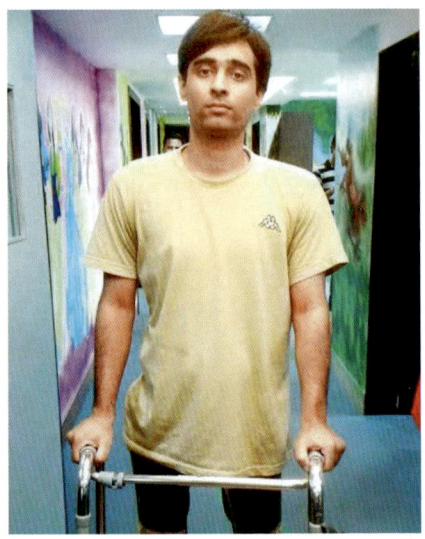

Major S.S. Shekhawat on Mount Everest, which he has summitted thrice

Major Manish Singh during his treatment, after getting critically shot in an army operation in 2012

Army wives, Sujata Devi and Vinita Devi, and army mothers, Bhagwati Devi, Indira Devi and Shashikala Devi, of operatives who made the supreme sacrifice in Operation Rangdori Baihk in 2020

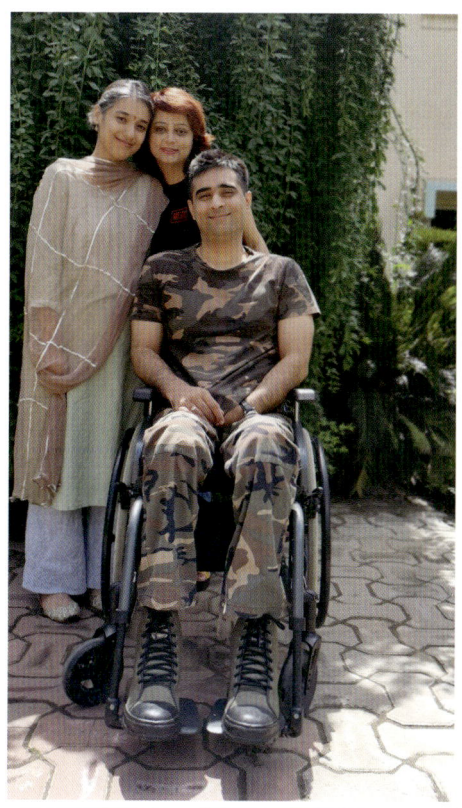

Major Manish Singh and his wife Aastha Panwar with the author

Captain Tushar Mahajan's parents and the author with his bust in the background in Udhampur in 2021

Captain Tushar Mahajan's bereaved mother paying a floral tribute on her son's tiranga-wrapped coffin in 2016

Colonel Santosh Mahadik's funeral in Pogarwadi, which was attended by then Defence Minister Manohar Parrikar, several other dignitaries and by people from all over the country

A young Tushar Mahajan in NDA 6th term as Divisional Cadet Captain

A young S.S. Shekhawat

Mahendra saab's grit, valour and mental alertness cannot be defined merely by the two medals he's awarded with, whether it is the Kirti Chakra or the Sena Medal that he earned for a covert nerve-wracking operation which is still classified and cannot be written about for the sake of national security and diplomatic constraints.

His senior officers, or fellow JCOs, who have served along with him say that if valour is to be judged by the number of medals you earn, then Mahendra saab deserves to have earned many more. He is like a mountain for the legendary 9 Para (SF)—standing tall in his tenacity and resilience. With his large frame, he dwarfs others even in the wheelchair on which he now spends his days.

The 2IC of 9 Para (SF) has had the honour of serving with Mahendra saab. He tells me, 'Ma'am, throughout his illustrious career, which spans over thirty years, he has served in the Valley and has participated in all the major operations that the 9 Para (SF) is known for. He has not just survived them all, he has conquered them all. The best part about working with him is that he is a true leader. He likes to carry the whole team with him. He knows his skills well. We were all heavily dependent on him.'

When I tell Mahendra saab of his unit's admiration for him, he smiles shyly and tells me, 'There were times when I thought of giving up, but the training of a Special Forces operative always got me through. The training emphasizes on making a soldier tough, more mentally than physically. If things don't work out as planned, we need to be calm and apply every cell in our brain towards controlling the damage, which is what I do these days. I remember the condition I was in after my last operation. People thought I would eventually die, then they thought I would never be able to get up and walk, that I would spend the rest of my days bedridden, like a vegetable. But here I am, sitting and narrating my story to you.'

When I heard about Mahendra saab, I knew right away that here was a man whose glories had remained largely unsung throughout his life in service. This, despite his unimaginable courage and contributions to the safety, sovereignty and security of this nation. Mahendra saab is a true legend. It's time we bring forward stories of more legends who have served the nation like Mahatma Gandhi, Khudiram Bose or Sukhdev did. We need to give new superheroes to the next generation. The onus is on the storytellers.

~

January 1992
Pulwama
South Kashmir

A 2.5-ton Army truck was speeding on the road in the wee hours of the morning. In the back was a fighting squad of six commandos, with the driver and an officer in the front. The squad was on duty carrying out a cordon-and-search mission. It was a time when militancy was at its peak and the exodus of the Kashmiri Pandits was an unfortunate reality. Unrest prevailed across the Valley. Every day heralded fresh blasts, firefights, encounters, disappearances, arrests, murders and gang rapes.

Amidst all this turmoil revolving around politics, Islamic radicalization and insurgency, the commandos of the 9 Para were pulled out of Operation Pawan in Sri Lanka and deployed instead as the first specialized commando unit in the Kashmir Valley—as part of the newly launched Operation Rakshak.[13]

The men in the back of the Army truck, then, were the guardians of the nation—a part of the extraordinary league we call the Indian Army. The Army had started taking control of the situation as soon

---

13 Operation Rakshak is an ongoing counterinsurgency and counterterrorism operation established during the height of insurgency in Jammu and Kashmir in the 1990s.

as the civil administration crumbled, before a well-planned agenda of an enemy nation, whose goal was separating Kashmir from the rest of India, could be fulfilled.

There were no guidelines then, no fixed protocols. Many locals were filled with hatred against India. The mosques would play *'Aey jabiro, aey zalimo, Kashmir hamara chhor do. Hai haq hamara aazadi* (O tyrants, O cruel oppressors, leave our Kashmir. Freedom is our right!)', asking Indian security forces to leave the lands that the locals were so desperate to hand over to Pakistan. Stone pelting was the norm, and military trucks sped up to 100 km/hour while passing through villages.

All this was too much for eighteen-year-old Paratrooper Mahendra, who had just joined the unit. Until now, he had been a part of the military convoys which arrived at the overflowing camps of Kashmiri Pandits in Jammu to hand out rations and relief supplies. The memories of the heart-wrenching cries, disease, injuries, the starved young children howling in the laps of their mothers and the old fathers looking for lost sons would steal his sleep for a long time. His mind would be endlessly disturbed as he tried to solve the puzzles of the massacre which was happening right before his eyes. This young man, who belonged to a peaceful village in Rajasthan, had only experienced peace and fraternity growing up. He found himself incapable of processing these horrors. Why were people being forced to leave their land and their homes? He would often think about how unfair everything was.

That day would not be so different. There had been an incident of communal disharmony in a village. There were reports of militants present in the village. The Troop Commander had asked the troops to board the Army truck and carry out a cordon-and-search mission before dawn broke, to avoid any aggressive stone-pelters.

The truck was speeding towards the village, which they were just about to reach. Mahendra had bent down to quickly retie his shoelaces before he jumped down from the truck. But before he could do anything, a loud blast rocked the truck off the road.

They would later learn that the truck had sped over an improvised explosive device (IED) embedded in the road. The officer in the front seat had been blown to pieces. There was blood everywhere, and the cries and shouts of his fellow soldiers echoed loudly in the air. For young Mahendra, it was a first-hard taste of the consequences of terrorism. The grotesque scene left an indelible impression on his mind.

The survivors were eventually evacuated by a nearby unit. It was at this point that Mahendra silently vowed to fight terrorists till his last breath.

During those moments of despair, Mahendra also found himself thinking of his young wife, Savitri, whom he had left back in his village. All three brothers had married the three sisters of one family. They all lived amicably as one joint family—everyone looked after each other. Because of this, Paratrooper Mahendra was able to focus so well on his job and training.

However, at that point, as he waited to be evacuated, he remembered 16 May 1990, just three months before his journey of becoming a paratrooper began. That was the date of his wedding to Savitri. He thought of how vulnerable, shy and pretty she was, with her face always covered demurely in a veil. She was a trusting girl who believed in everything he said and every dream he had. For Mahendra, Savitri was his Lady Luck. From the day she had stepped across the threshold of his home as his bride, he seemed to have good fortune. He had joined the Army within a few months of marrying Savitri.

Even though he was far away from his family, Mahendra realized that his safety was paramount for his family. His young wife awaited his return every day. Until now, even while he had been part of active operations, Mahendra had never encountered any terrorists. As he lay recovering in his hospital bed, he wished desperately to be well enough to raise his gun against every terrorist.

~

April 1992
Somewhere in Baramulla
North Kashmir

Paratrooper Mahendra was part of legendary Capt Arun Singh Jasrotia's Alpha team. Jasrotia was his Troop Commander, and that thought itself kept filling Mahendra's heart with great pride. Capt Jasrotia had earned a legendary name for himself in the world of the SF. He was among the first lot of a new generation of officers of the Commando Battalion (later converted into the SF in 1994) that was fighting against the new wave of terrorism engineered by Pakistan across Kashmir.

Capt Jasrotia had been part of many successful operations against local and foreign terrorists when insurgency had erupted across the Valley. Many locals had started training in camps across the LoC, and infiltration had begun into Indian territory by trained terrorists. It was a new situation for Indian security forces and the Indian government. Had it not been for daredevil officers like Capt Jasrotia, SM: AC (posthumous), and many other Indian soldiers, gaining a foothold in the fight against terrorism would have been challenging, to say the least.

Honorary Capt Mahendra Singh told me, 'When I look back, I feel truly fortunate to have been under the wing of such legendary officers like Capt Jasrotia saab and Maj Sudhir Walia saab at the start of my military career. It is a rare honour, I tell you. Their jungle- and mountain-warfare skills were extraordinary, and I learned many of those skills from the legends themselves—from laying an ambush in rough terrain in the mountains to surviving in the jungle. I remember my very first contact happened when I survived a bullet by a mere inch, when I was a part of Jasrotia saab's Alpha team.'

In those days, Afghan terrorists were rerouted by Pakistan to India to train and support local militants. The Afghan terrorists were high on Inter-Services Intelligence (ISI) money, and they played a crucial role in instigating insurgency in the Valley. It was then that the Indian Army had the idea of establishing proper SF units at large and began modernizing the Army seriously.

The battalion now started adapting to a new enemy, learning their customs, language, skills and even matching their weapons. Fighting militants as militants was the new motto.

The inputs were discreet, and it was reported that two Afghan terrorists and three local terrorists from the Jammu Kashmir Liberation Front (JKLF) had taken refuge in a village in Baramulla.

The 9 Para (Commando) gave the task to the Alpha team led by Capt Jasrotia, then a Troop Commander. He divided his squads into two parties: a cordon party and a search party. The cordon party was further divided into inner and outer cordon parties. Their job was to lay an ambush inside and outside the village. The search party, on the other hand, was responsible for searching the houses once the cordon had been established.

The team laid out a brilliant ambush in the night, fully aware of the chances of a firefight breaking out. It was at first light that the commando squad sighted five people who looked like local villagers walking towards the outer ring of the village. They looked harmless enough and were even dressed in attire similar to what the locals wore on an everyday basis. The team held its fire for a while, watching them. But Capt Jasrotia sensed instinctively that something was wrong and accordingly sent a message over the radio. They were, for one, walking in a near-perfect tactical formation, maintaining a distance of 5–10 metres between each other. If they had been villagers, they would have been walking together. They were definitely terrorists.

The team started firing at the men. Paratrooper Mahendra was beside Capt Jasrotia, who was also firing relentlessly, fuelled by the months of anger and frustration of watching helpless families fleeing their homes and comrades dying in the fight against militancy. He did not even realize that, in his adrenaline rush, he had compromised his cover. He was now fully exposed, but he kept on firing. It was only then that he felt the soil on his face and heard Jasrotia's voice shouting, 'What the hell are you doing out of your cover? You have to kill them, not get killed. Get back, get back!'

Coming to his senses, Paratrooper Mahendra immediately jumped back to his cover, adjusted his posture and resumed firing. He realized later that he had been saved just by an inch because the bullets had hit the ground beneath his feet, hitting his face with just dirt. It had been his lucky day. He was clearly destined for greater things in life.

Three terrorists had been successfully eliminated during this initial firefight, while two took refuge in a nearby house. There was an elderly man and several women in the house. The team found itself dealing with a hostage situation. Bullets were useless here, and grenades or rocket launchers could not be used. It took several rounds of negotiations to convince the terrorists to release the hostages by noon. Capt Jasrotia successfully convinced them of the value of a peaceful surrender, and the squad started preparing for the same.

However, they didn't expect that the doors of the house would immediately close soon after releasing the hostages, and neither did they expect the terrorists to start firing again. Later, they would learn that the terrorists had orders from their Pakistani masters to fight until their last breath, in the name of Allah, rather than surrender to the kafirs on the Indian side of the border. Eventually, Capt Jasrotia launched grenades into the house, setting the house on fire and eliminating the terrorists. The entire operation was wrapped up by late evening that day.

The operation was a great success with zero casualties on the side of the forces. The encounter made headlines for several days—a stern warning to militants that no mercy would be shown to them if they threatened the sovereignty of India.

While it was just another day and another operation for Capt Arun Singh Jasrotia, it was a monumental lesson for Paratrooper Mahendra, who vowed never again to lose his senses and always control his nerves during future operations. He knew now that death was always a literal inch away.

~

January 1993
INS Venduruthy
Navy Base, Willingdon Island
Kochi
Kerala

Raiding ambushes, tactical training and jungle raids were a regular part of young Mahendra's life. He was a quick learner. He had developed the amazing knack of correctly reading the terrain and was always volunteering to act as a scout in all operations. He was also an excellent firer of crew-served weapons like rocket launchers and machine guns. Due to his multifarious talents, he was chosen by his unit to participate in the Army Combat Underwater Diving Course at INS, Kochi, considered to be one of the most challenging courses in the world of the SF.

The young man from the desert lands of Rajasthan was thrilled. He had a natural affinity for the water, even though he had heard that qualifying for the course could be challenging. One needed to be both mentally agile and physically fit. When Paratrooper Mahendra stood with a hundred volunteers from all the arms of the Indian Armed Forces for the screening test at INS Venduruthy, Willingdon Island,[14] everything looked and felt like a dream to him. Everyone seemed to be more qualified than he was. There were Indian Army officers, Navy sailors, executive officers of the Navy and even civilian personnel. They looked experienced and confident, but as soon as the Navy instructor told them to run, Mahendra knew that he still had a chance. After all, running was his forte.

The screening test lasted fifteen days, comprising various activities from swimming to running. After each test, some boys would be sent back. Eventually, fourteen boys out of the hundreds of applicants were announced to be eligible for the course. Paratrooper Mahendra Singh was one of them.

For the next three months, he underwent rigorous training, with the chances of being sent back increasing with each new step—but he made it through. From basic diving training, swimming or floating with a diver set to advanced combat diving training of bombing ships or jumping from sky-high bridges, Mahendra excelled in everything. When the brass diving badge that depicts a diver geared with combat loads and carrying the Ashoka emblem on his back was pinned onto his uniform by the Commodore of INS Venduruthy, certifying him as a qualified combat diver ready to carry out special underwater operations, it was the happiest he had felt in a long time.

He could not wait to share this news with Savitri, who was another key to his happiness. When he had last spoken to her, she was pregnant with their first child. He had not been home for a long time. From there, he had gone to Kochi for the diving course. But now he had a window to visit his family.

---

14 https://www.indiannavy.nic.in/content/ins-venduruthy-seamen-training-establishment

When he called home, his elder brother Havildar Ramvilas, who looked after their joint family, informed him that Savitri had delivered a healthy baby boy on 15 April 1993. The happy news transported Mahendra to cloud nine, and he left for his village immediately. When he held his baby in his arms, his joy knew no bounds. The nation still came first. But now he had more reasons to fight. He named his firstborn Pradeep, which means 'light'.

~

15 September 1995
Lolab Valley
North Kashmir

Insurgency was at its peak in the mid-1990s. By now, Paratrooper Mahendra had become an expert in counterinsurgency operations. Every day there was something new to learn, new challenges as well as heartbreaks. Mahendra saab shared with me his pain at losing Capt Arun Singh Jasrotia, SM: AC (posthumous), on the fateful day of 15 September 1995 in the Lolab Valley Operation. While it taught him a lot, it still haunts him.

He said, 'Madam, I could not eat properly for a long time. We are in a profession where the risk to comrades' lives is always there—but trust me, the pain is always real, no matter how many comrades we lose. I have always been a part of many risky operations carried out by the legends of 9 Para (SF), from Jasrotia saab to Walia saab—and losing them has been so heartbreaking. Sometimes, I think the bullet should have hit me instead of them, but then this is how soldiers embrace the most glorious deaths. They sacrifice their lives so that their valour can inspire an entire generation of youngsters to never hesitate to make sacrifices for the nation.'

~

1996–99
The National Security Guard
New Delhi

Pitch-black dungarees clothe his tall frame; a black mask over his face only reveals his eyes. Weapons—the 9x19mm Parabellum submachine gun Heckler & Koch MP5 clutched between his hands and a Glock pistol tucked in his belt—add lethality to his aura. The upper half of his black dungarees is studded with pockets: small, big, rectangular and square, each carrying various utility items from grenades to James Bond-style covert spy devices, and his trusty radio set.

It is tough to imagine 9 Para Special Forces operative Mahendra Singh behind that mask, as he stands guard at Palam Airport (now known as Indira Gandhi International Airport) attentively. It has been several years since Paratrooper Mahendra was posted in the Kashmir Valley, so his deputation to the National Security Guard (NSG) felt like a welcome change. He had lost too many friends and mentors in the fire of terrorism in Kashmir, which was at its peak in the 1990s.

Mahendra Singh was assigned to 52 Special Action Group[15] and tasked with carrying out anti-hijacking operations. It was not easy—this was one of the most arduous commando trainings in the world of the SF.

There were rigorous battle-obstacle courses, an intense Battle Physical Efficiency Test, physical proficiency tests, physical training, parades, unarmed combat techniques, judo, karate, and many more exercises which made up an average day for all NSG trainees.

---

15 Unit of Special Action group of NSG trained and equipped for counter-hijack operations and sky marshal duties. Popularly called Hijack Busters.

As Mahendra saab explained, firing is given utmost importance while training an NSG operative. Their aim should be accurate because, whether it is a hostage-rescue operation or anti-hijacking scenarios, the presence of civilians demands the utmost precision and care. As a result, operatives get only one chance to safeguard the lives of innocents. So, a prospective NSG operative is trained heavily in shooting and sniping for hours, long- and short-range firing in closed rooms, and more.

During this advanced training, Mahendra saab also learned about India's aircrafts—especially how to raid them and rescue the hostages. It was a thrilling experience for him as he understood how to wield the bundles of ropes over his shoulders used to climb on to the aircraft and the various other gadgets and advanced weapons.

Paratrooper Mahendra soon became an expert in anti-hijacking operations, and he had also achieved a black belt in karate by then. He was posted to various airports to ensure their safety and often travelled as an undercover operative to perform sky marshal duties in airplanes as part of his job. His NSG tenure also allowed him to spend quality time with his family. He became a father for the second time when his daughter, Suman, was born on 5 January 1996. It felt as if everything in his life had fallen into place. His family was complete, and they could holiday together and live an ordinary but peaceful life.

Now that Mahendra had built up some savings too, he could invest in his village along with his elder brother. He bought some farming land and constructed a better house. It was everything he had ever dreamed of. But by the time his deputation ended, the Kargil War had started. Mahendra was itching to rejoin his unit. He knew Mother India needed him.

~

July 1999
Kargil War
Sando Top

Lance Naik Mahendra Singh was part of the 9 Para (SF) team on the quest to reoccupy Sando Top, a steep and dominating peak in the area, which was strategically of immense importance to India.

Though the Kargil War was a bitter three-month full-blown war mainly fought by infantry battalions with support from artillery battalions, eight out of the ten battalions of the Para regiment took part in operations to push out the intruders.

The 9 Para (SF) actively participated in the battle in the Drass–Mushkoh subsector, and also led the assault on Sando Top, of which Mahendra was a part, and fought the battle against enemies along with his comrades. There was a lack of input regarding the operation, and the unit had to act on minimal information. There was hardly any information about enemy logistics—the type of weapons used, their backup lines and supply routes. Still, the unit did not back down.

Eventually, though, the operation was successful and the heights were captured.

Lance Naik Mahendra was part of the Alpha team which was actively involved in capturing the height with full frontal assaults. He also operated under Maj Sudhir Walia, SM*: AC (posthumous), for a few operations in the war.

Mahendra saab shared with me, 'Walia Saab had obtained special permission from the Chief of the Army Staff Gen Ved Malik during his appointment to rejoin his unit on the battlefield at Kargil. Within ten days, he had taken over the Alpha team and our josh was at an all-time high. One became fearless just by being in his company.'

Lance Naik Mahendra Singh performed many surveillance and reconnaissance operations as part of the 9 Para (SF) battalion. During those operations, he discovered Chinese Claymore mines as well as tracks high up in the mountains evidently carved by explosives, even full fiber bunkers with a good amount of ration and basic amenities —a clear indication of a full season of hard work by a large labour force. The 9 Para (SF) eventually accomplished every given target, and Lance Naik Mahendra was part of this successful team. It was a great learning experience for him, one that would pay good dividends in the times to come.

The Indian Army's quest for peace did not stop there. Instead, it intensified after the Kargil War. The defeated Pakistan intensified infiltration from across the border. Intense counterinsurgency operations were followed after the Kargil War in Jammu and Kashmir. Insurgency levels were so high that it took a division-sized force comprising helicopters and guns to reassert peace in the area.[16]

Operation Sarp Vinash[17] was put into action to initiate encounters with terrorists. The Army units would move out of their bases looking extensively for militants. They lay effective cordons on the way, climbed for ten to fifteen hours in the mountains, laid ambushes, and encountered terrorists almost daily. Lance Naik Mahendra Singh was a Squad Commander by then, and his skills and tactical acumen can be gauged by the simple fact that he never lost a man under his command, despite being part of dozens of contacts.[18]

~

16 Lieutenant General P.C. Katoch, *India's Special Forces: History and Future of Special Forces* (New Delhi: Vij Books, 2014).

17 Praveen Swami, 'Operation Sarp Vinash', *Outlook*, 27 January 2022, https://www.outlookindia.com/website/story/operation-sarp-vinash/220331.

18 The firefight with terrorists by security forces is called contact.

28–29 August 1999
The Haphruda forest
Kupwara
Kashmir

Just a few months after the Kargil War, Lance Naik Mahendra had the rare honour of participating in Maj Sudhir Walia's last operation. This mission was one of a kind—an explosive display of raw courage and audacity that tore the enemy apart. Call it fate, the alignment of the stars or simply harsh reality but, by now, Mahendra saab had been part of all the legendary operations of the 9 Para (SF) and had worked closely with the most audacious names of the battalion. He is wheelchair-bound today, but it is no less than a miracle that he lived to tell the most legendary tales of human grit and valour.

Lance Naik Mahendra has served as Maj Sudhir Walia's radio operator and scout. Maj Walia was an icon for young Mahendra.

Mahendra Saab recalled, 'I was Walia Saab's scout most of the times. We shared a great rapport. I loved being his eyes and ears, and he relied greatly on me. We killed five terrorists in 1995 in the same area where he eventually made the supreme sacrifice in 1999. Call it fate, but the then SM [Subedar Major] appointed me to another team and appointed Sepoy Kheem Singh as Walia Saab's scout during his last operation. Eventually, Naik Kheem Singh[19] earned the great honour of sacrificing his life along with Walia Saab, while all I could do was listen to the radio conversation during the same operation in another area. I could not reach my brothers on time, and I will always regret it.'

Lance Naik Mahendra was a part of the same intelligence-based operation but with another team operating at a distance from Maj

---

19 General V.P. Malik, *Kargil: From Surprise to Victory* (New Delhi: HarperCollins, 2010), p. 188.

Walia's team. The task was to raid militant hideouts in the Haphruda forest in Kupwara. Maj Sudhir Walia, SM*, was awarded the Ashoka Chakra posthumously—the peacetime equivalent of the Param Vir Chakra.

His voice choking, Mahendra saab told me, 'Madam, I still wonder why I was not with him. I should have been there with him.'

The moment turned sombre and I could not complete my scheduled interview with Mahendra saab that day. Talking to someone who has lived through those times and reminding him of those dark memories is perhaps the most challenging task on this planet, and it was during moments like these that I hated my job.

~

December 2008
Somewhere in Baramulla
Jammu and Kashmir

Mahendra saab had always been in the field of counterinsurgency and has undertaken numerous encounters throughout his military career. It is impossible for me to compile all of them. But I thought of elaborating on one unique daredevil operation undertaken when he was a havildar.

The Military Intelligence was catching signals from a group of terrorists from the forests of Baramulla district, but the frequency was never static. When the message was relayed, several parties of the Rashtriya Rifles moved out to look for the contacts of these terrorists. However, two months passed in the snow-covered forest, and they still had not been able to locate the hideout. What was astonishing was that not even footprints were found in those two months. However, the Indian Army's high-tech communication devices affirmed the continuous presence of terrorists in the area, from where they were comfortably operating. This was a troublesome situation. If not eliminated, they could create havoc.

Finally, the 9 Para (SF) team was informed, and twenty-four troops embarked on the operation. The terrain was difficult. A snowy forest of thick coniferous trees with steep gradients, sharp inclines and ferocious nallahs made daily patrolling difficult. The Rashtriya Rifles team had been doing daily patrols for the past two months in that terrain, but there had been no output.

It was at this point that the battle-hardened and experienced Havildar Mahendra Singh came up with a unique plan. The 9 Para (SF) team decided to disguise themselves as militants and live in the area in hideouts under trees or caves. It was a complex task, but it was also the most feasible option under the circumstances.

And so, Mahendra saab and his troop commander looked out for dhoks[20] in the higher reaches of the forest.

Mahendra saab and his men soon found suitable dhoks in the forest area and established an excellent harbour meant to be a temporary control base from where the troops could replenish, rejuvenate and rest, and search parties could be sent out further ahead in the jungle.

Havildar Mahendra Singh also divided the troops into four parties, which would go ahead into the deep forest around the most likely coordinates and stay there for some days. At intervals, a new party would replace the fatigued forward party and continue with the surveillance impeccably. All of this modus operandi was a picture-perfect illustration of some of the world's finest military tactics, under the impeccable leadership of Havildar Mahendra Singh.

---

20 Dhoks are huts made of mud, stones and dry branches and are used during the summertime by shepherds and farmers. During winter, such dhoks on the higher reaches of forests are all empty because the farmers return to their villages. So these dhoks serve as good shelter houses for force and terrorists both.

While staying in the jungle, the men realized that the terrorists cleverly maintained their radio communication with their bosses in Pakistan or local OGWs in the Valley only when snowfall happened, and communicated only during random strolls. In this way, they could ensure their coordinates would always be random, with the snow serving to eliminate their footprints.

As for the troops, they faced a lot of difficulties. They would hide behind and under trees and conduct surveillance—creating little spaces by moving snow around the roots of the tree and eating whatever little stale food they had with minimal movement. The weather in December was at its cruellest too. There was heavy snowfall and almost zero visibility.

Havildar Mahendra strived to keep the morale of his troops high. It was his duty to bring the squad back alive. Despite walking for days to establish contact with his search party, this resilient soldier never once lost his calm. He often cracked jokes about how terrorists had a knack for operating under difficult weather conditions. Sometimes, he went to the back of the line to give water to the last soldier, or to share the heavy load of a fatigued fellow soldier. There was a reason he was known affectionately as a great mountain. After three days of continuous recce and surveillance, the team finally saw smoke filtering out of a cave covered with snow, while a sentry stood guard outside. A firefight began, and the sentry was killed on the spot. The second terrorist, named Haris, who had been talking to his girlfriend, ran and hid in the jungle. The commandos started looking for him, and a desperate a cat-and-mouse game ensued.

Haris[21] was the terrorist commander, and he had fled along with two of his men. Later, he would ask them treacherously to engage the troops in a firefight so that he could escape. It was a suicide mission—perhaps Haris knew it too. Two terrorists were killed on

---

21 The name given on his Pakistani passport

the spot. Haris managed to escape, but it would only be a brief period of freedom, because he was eventually killed by a Rashtriya Rifles battalion at the outside cordon.

When the hideout was searched, an immense amount of ration, cash, weapons and ammunition, sufficient for months, was found. Clearly, the terrorists had formulated an excellent long-term plan to operate in the area for months.

Later, the operative narrating the story told me, 'Ma'am, we have seen these terrorists up close and personal. Some try to cause chaos before embracing their end, but most run for their lives when the Indian Army enters the arena. This jihad business looks good only on social media or their websites. But in reality, they don't hesitate to sacrifice the life of a friend instead of their own. One fact remains—the moment they step on to Indian soil, their death is certain. These terrorists infiltrating across the borders are the mere guinea pigs of our enemies. Their end is always sad.'

Subedar Maj/Honorary Capt Mahendra Singh, KC, SM, has had a long and illustrious career and has given his all for every posting he has held—from his exemplary service in 2005 as a havildar in Ethiopia and Eritrea as part of the UN peacekeeping force to his innovative methods of teaching when he was a diving instructor to the probationers of the 9 Para (SF) battalion. An expert in counterinsurgency operations, including covert operations, he has also been part of numerous operations along the LoC. For his stellar role in these operations, he was awarded the Sena Medal on Independence Day in 2013. Mahendra saab is not just a soldier. He is also a hero whose legendary career spans decades.

When I visited the 9 Para (SF) unit in the hope of finding some anecdotes about Mahendra saab, I was nervous. I did not know if I would be able to cover everything his comrades were telling me in just a few pages and be able to justify his stature in the community, his sacrifices and his towering personality. So many highly decorated officers—not just jawans or JCOs—came forward to discuss their association with Mahendra saab. In my journey as a defence author, I find it difficult to compile stories of other ranks and JCOs because of the lack of extensive support and material my stories require. This instance highlights the kind of bonhomie and bonds officers and jawans of the SF battalions share together.

The current SM[22] saab of 9 Para (SF) fondly remembered their joint Army–Navy commando exercise in Mumbai. There was a stiff competition between the Army and the Navy to score the maximum points and beat the other. He narrated how tricky the particular exercise was: the participants were left on an island with their rations cut off, and they had to escape from the island only to raid an airport. Each Army and Navy team had their respective sets of officers: JCOs and non-commissioned officers, but eventually, the Army team, under the capable leadership of Mahendra saab, won that particular exercise.

When I asked Mahendra saab about it, he laughed and said, 'Yes, I remember it happened in 2007. In the water, Navy commandos are better than us. But when it comes to land warfare—whether it is running through the jungles, slithering down from helicopters or raiding airports—no one can beat the Para Special Forces operatives. We are also used to operating in jungles with minimal rations, so it did not deter us when our rations were cut off.'

He continued, chuckling, 'I remember this particular exercise because the locals on that island found our activities suspicious and reported us to the local police! So, in the middle of the night, the local police surrounded us in boats and asked us to drop our

---

22 Subedar Major of the unit, who is the highest rank amongst JCOs. The Commanding Officers commands the unit but it is the SM of the unit who can be called its backbone, acting as a cohesive force between officers and Jawans. A revered figure in the unit.

weapons and surrender. What followed was a series of confusing events. It also was reported in the press. Eventually, the Army and Navy headquarters had to intervene. Senior police officials had to inquire how this whole thing had gone so out of hand, because such exercises are always carried out after informing the civil authorities. Still, in the end, when we boarded our flight back to the unit, there was a riot of laughter in the plane.'

I smiled. Real-life superheroes don't look like Superman or Batman. They look exactly like Mahendra saab—a man who has lived a full life, which, when revealed, looks like a bundle of adventures and glory. He is one who lost everything in service to the nation but still sings its praises.

Mahendra saab lost both his legs. He is wheelchair-bound, but that does not deter him from loving his country. When you talk to him, his infectious good energy instils courage in you and gives you the hope that as long as there are men like him, we can sleep peacefully in our houses. Our economy will flourish, the country will develop. Men and women will be free to chase their dreams. All because men like Mahendra saab guard our borders day and night.

I remember the then 2IC of 9 Para (SF),[23] who had the privilege of serving alongside Mahendra saab, first as a Troop Commander and then as his Team Commander. He used to say that Mahendra saab was their guiding light, no matter what capacity he was serving in. He remembers how Mahendra saab saved his life once during a snowbound operation. The officer also explained how innovative saab had been in his ways, whether it was operations in the jungle or training exercises imparted to young probationers—he would always think out of the box.'

---

23 Name of interviewee and date of interview withheld to maintain the privacy of the serving personnel.

Naib Subedar Mahendra Singh was tasked with designing a new course for probationers for conducting raiding and ambush exercises. Since the SF are all about unconventional warfare techniques, the challenge was to beat the terrorists in their comfort zone—from deep jungles to the sharp nallahs which they usually used to escape.

Mahendra saab, who had just become a JCO in 2009, came up with a capsule course where probationers were taught to sit quietly underwater with the help of a tube while waiting for their target. Someone above the water would signal the underwater operative with a fixed set of instructions. When the target came closer, the operative would spring out of the water and attack. The water would be muddy and cold, and operatives ran the risk of hypothermia. But Mahendra saab kept jumping in and out of the water, shouting instructions and boosting everyone's morale. Eventually, that technique also proved beneficial for the unit during actual operations in hilly terrain, near a pond or a waterfall.

Anecdotes of dealing with wild dogs and even cheetahs in the jungle during raids or ambushes abound about saab. Then there are the tales of inter-company competitions during the annual war exercises and the times when saab went out of his way to help his comrades, even risking his life selflessly for the '*Naam, Namak, Nishan*' he so fiercely believed in. Some tales are funny, some heart-warming and bright, while some seem simply mythical. It was hard to believe what one man could achieve through sheer grit, resilience and training.

During his last operation at the Haphruda forest, Mahendra saab sustained a gunshot wound in the abdomen, with a fracture in his spine and intestines, in his quest to save his comrades' life. When he was shot, despite being in unbelievable pain, he only wished for the safety of his men. His elder brother Havildar Ramvilas (retd), who was with Mahendra during his treatment at the Army Research and Referral Hospital in New Delhi, told me that even when he was fighting for his life, all Mahendra cared about was his comrades.

Ramvilasji told me, 'Beta, he was unconscious for days. His spine was exposed. We could see it, there was pus filled in it. And his intestine was also ruptured—this resulted in faecal matter spilling out in the body. It was an alarming situation for months because both his upper and lower flanks were exposed. He spent days lying on one side only. He was in a coma, too. It was a critical situation. The doctors did not believe that he would live. But all Mahendra would do in his half-conscious state was call out the name of Lance Naik Mohan Nath Goswami. He would shout in the middle of the night, "Where is Goswami? Where is my chap? Is he all right?" Eventually, the doctors had to tie his hands and legs to keep him still. This went on until he recovered his memory and doctors declared that he would survive.'

When I talked to his doctors, they unanimously agreed that they had never seen a tougher guy than Mahendra saab. They told me that because it was a bullet wound that had caused such serious injuries, they had to keep the organs exposed. Many a time, they cut the skin from the area or even stitch up the area without giving him anaesthesia. But never once did Mahendra saab stop fighting. If he is alive today, the doctors said, it is because of his grit and will to live. I left wondering how we can ever repay someone like Mahendra saab. No medal justifies his sacrifices.

Mahendra saab was in the hospital for two long years. It was only after this period that he was able to visit his village. As I wrote earlier, people from not only his village but from many villages across Jhunjhunu arrived to get a glimpse of this braveheart. Rajasthan is a land of veers[24]—its people recognize true courage. Almost every family has at least one person in the armed forces.

---

24 Bravehearts

Mahendra saab has made them proud. He was awarded the Kriti Chakra, the second highest peacetime gallantry award in 2016, the only one awarded to 9 Para (SF) till date and, as much I know, the only one awarded to someone from Jhunjhunu district. The then President of India himself stepped down during the ceremony to pin the medal onto Mahendra saab's uniform, and all the ministers present greeted him warmly. The applause wouldn't stop as his story was being narrated during the felicitation ceremony at Rashtrapati Bhavan. Saab was still in the hospital at the time, and he mentioned how kindly the Army provided him a special bed-cum-wheelchair which he was harnessed onto with belts so he could sit up properly for the once-in-a-lifetime ceremony.

Mahendra saab also mentions that he never felt abandoned or alone because of the support of his unit,[25] which ran pillar to post to provide the best of everything for him. They took care of the treatment, the finances and even ensured that after his treatment he would be attached to the station headquarters, from where he retired with dignity and pride in March 2021.

Today, Mahendra saab stays with his family. Both his children are well settled. His son, Pradeep Kumar, completed his engineering in 2015 and now works as an assistant manager with the Reserve Bank of India, and his daughter, Capt Suman, completed her MBBS in 2020 and carrying forward the legacy of the family by serving in the Army as a doctor. The credit of raising his children so well goes to the simple and shy Savitri, who herself had been uneducated but ensured that even in the absence of the father, their children got the best education. This is the strength and tenacity of an Army wife; the

---

25 The way 9 Para (SF) looks after their men is commendable, especially with regard to those killed or disabled in action. The commanding officers take it personally upon themselves to ensure the best for the men and their families.

force behind the forces,[26] protecting everything they leave behind. Today, Savitri is delighted to have her husband home and safe after thirty years of service, but Mahendra saab still misses his unit life.

He misses the routine and rhythm of his beloved 9 Para (SF), where he made a life with his comrades. His son says that when his Papa was in service, they would never travel, shop or do any of the things that average families did. His father never attended his parent–teacher meetings—nor did Savitri, who was shy and burdened with housework in Mahendra's absence.

But as Pradeep and Suman grew older, they understood what it really meant to have a loved one working in the SF. They watched their mother crying or worrying about their father's safety. Sometimes, they didn't hear from their father for months at a time. All this made both children grow up well before their time. So, naturally, having him back home with them permanently is pure bliss.

Mahendra saab is very proud of his family and has the deepest gratitude for his wife, who never gave up on him. Even today, her entire day revolves around him. She tends to him with the utmost care, now that he is in a wheelchair. The two are making up for lost time. If you ever stopped believing in true love, I hope you now know it exists, and it is indeed kind, strong and selfless.

When I asked Mahendra saab what he now looks forward to most in life, he replied, 'Madam, I have lived my life to the fullest. I have done and witnessed things not all men have the privilege to do. My children have also made me proud, even when I was never around for them. They are the best children in the world. There is nothing that I want now except to be able to walk again or at least be able

---

26 The author's previous book, *The Force Behind the Forces* (Penguin Random House, 2021), provides a deep insight into the courage and resilience of Army wives and highlights their contribution to both the army and the nation.

to do my job myself one day. There is one more operation pending and many physiotherapies going on, so at this point, all I am focused on is getting better. I have everything else, including the love of my family and my unit.'

When asked if he had any message to share with the citizens of this nation, he smiled and said in his calm way, 'A soldier only seeks pride and dignity. Whenever a soldier or his family visits any civilian for work, they must be given priority, and their problems must be heard and solved by any means. Not just the soldier but the contribution of his family must also be acknowledged by the people. This change in society would suffice. We only want respect.'

The story is based on interviews conducted with Sub Maj/Hon Capt Mahendra Singh, KC, SM, the protagonist of this story, his wife; his son, Pradeep Kumar; his brother Ramvilas; and hundreds of his comrades who have served with him over the decades. Many of these men are from various units and cannot be named due to security reasons, but I must give a special mention to the dynamic and dashing Commanding Officer and all the officers of the 9 Para (SF), who have not forgotten Mahendra saab's contribution and extended all possible help to me to tell his story. He holds a legendary status in his unit.

The story of Subedar Maj/Honorary Capt Mahendra Singh, KC, SM, is important because people need to know that our freedom and sovereignty is made possible because of the courage, contribution and sacrifice of men like him, who come from the heartland of India—from its small villages—and silently give their all. His story is just one among many of the unknown brave patriots who guard our nation every second with their every breath.

Jai Hind!

4

Captain Tushar Mahajan, SC (posthumous): The Enigma Who Went Too Soon

Somewhere in Bandipora
North Kashmir

EVERYONE IN THE remote village in the daaman[1] of a forest in Bandipora district were sleeping. From the nearby forest came the shrill chirping of crickets, the rushing noise of a babbling stream and the rustling of nocturnal animals that were waking from their morning's slumber. Snow-capped mountains surrounded the hamlet, dark and forbidding in the night.

It was at this unearthly hour that one of the villagers, Ahmed Dar,[2] heard desperate knocks on the door of his flimsy two-room house.

---

1 Foothills.
2 While the rest of the chapter is based on real incidents, creative liberties have been taken to write this particular scene fictionally to highlight the covert nature of SF operations braveheart Captain Tushar Mahajan was part of. These operations have not been made public considering their ongoing status.

He murmured in Kashmiri, complaining about uninvited people in the wee hours while opening the door, '*As-Salaam-Alaikum, chacha … Khairiyat* (Hello uncle, is everything okay)?'

A tall, fair, handsome man, with a long, flowing beard, dressed in a Pathani suit and skullcap, stood outside. He replied, '*Wa-Alaikum-Salaam, beta*, tell me, what is the urgency?'

Dar noticed the stranger's eloquent Punjabi Urdu accent—not spoken in Kashmir but in areas closer to Pakistan.

'*Kafir*[3] *toh nahi hai gaon mein* (Are there any Army men in the village)?' There was alertness in the man's calm voice, his eyes sharp in the moonlight.

'The kafirs had come in the afternoon but they left in the evening. Tell me, what can I do for you?' Dar said with great respect. He knew by now that this man was a foreign terrorist from Pakistan-occupied Kashmir (PoK). He had likely crossed the border by dodging the security forces in the jungle.

'Chacha, can I request you for roti and gosht? We have come from far away, and our rations are over. For the past two days, we have been starving,' the militant asked politely.

'Yes! Of course! Let me ask my wife to pack it for you. Fatima, do we have roti and gosht for our guests?' Dar shouted.

Soon, a pile of freshly baked bread and some leftover meat had been packed nicely into several polythene bags and handed over to the militant. He hurriedly accepted it, murmured shukriya and quickly walked back to join his friends, who were standing guard under the thick cover of trees with AK-47 machine guns in their hands.

Still standing at the door, the couple were able to take a good look at them. There were four of them, all clad in Pathani suits, carrying pitthus (backpacks) on their backs and several green pouches strapped around their bodies. They looked dishevelled, and their beards were long. Despite their lean frames, they looked exceptionally agile.

---

3 'Kafir' referred to the presence of the Indian Army in the area.

This kind of activity continued for four days—the militant would knock at unearthly hours on Ahmed Dar's door and ask for food, while the rest of his friends stood on alert with their AK-47 machine guns, taking cover around the house. As soon as they received the food from the couple, they would leave for the forest near the village.

A sense of ease began to develop between the villager and the militant. Maybe that was why Ahmed Dar invited the militant inside his house on the fifth night. The militant had refused the invitation that night. It was only the following night—the sixth consecutive day of their interaction—that the militant himself asked to enter Dar's home and share a meal with him. Dar immediately obliged, and cordially invited the other militants to join them too. Only two of them accepted his invitation—the last one preferred to stay outside on guard while his mates dined inside.

Inside, the house was decorated with typical Kashmiri papier-mâché. It was a modest home with minimal amenities. There were piles of blankets in one corner and posters praising Allah and jihad on the walls. The lady of the house, Fatima, offered the militants water in a bowl so they could wash their hands. Sitting on the earthen floor, the militants washed their hands and kept their guns down near them. They did not look like chatty men.

Dar ventured some small talk and said, *'Aur saathi kaise hai* (How are your friends)?'

He was referring to their masters in Pakistan. It had been a few days since he had been helping them, but he still knew nothing about their tanjeem.[4] He was curious.

The militant leader replied, *'Acche hain, chacha. Yahaan halaat kaisi hai* (They are fine, uncle. How is everything here)?' He was referring to the Army activities in the area. The rest of the militants hadn't spoken a word yet.

---

4 Terrorist organization.

Ahmed Dar replied, concerned, 'Halaat toh theek nahin hai. Kafir gashat lagte rehte hain. Pehle se zyaada laga rahe hain aajkal. Shayad aapke aane ki ittalah hai inko. Behtar hoga aap nikal jaaye (Everything's not all right. The Army keeps patrolling the area, much more than they used to do. Perhaps they have received word that you are coming. It would be best if you all leave immediately).'

The militant whispered, 'Hmm! Hum abhi aaye hain. Yahaan ka kuch pata nahin hai. Aap kuch madad kar sakte hai (Hmm. We've just arrived, so we don't know too much about the state of things here. Can you help us)?' All in eloquent Punjabi urdu.

Ahmed Dar replied, 'Haanji! Bilkul! Yeh hi to humara kaam hai. Aap kal raat ko gaon ke nambardar se mil lijiyega, main kal subah unko aapke aane ki itallah kar dunga—aapko sahi tanjeem se mila denge. Iss area mein patta bhi bager unki marzi ke nahin hilta. Hum sab kafiro ke khilaf iss jung mein unhi ke hidyat se apne farz ki tameel karte hai. Aap unse mil le, aapka kaam ho jayega (Of course. That's what I am here for. Tomorrow at night, I will take you to meet the village headman. I'll tell him in the morning that you will be coming to meet him. He will introduce you to the right people. In this area, not even the leaves in the trees move without his approval. Once you meet him, your work will be done).'

The food arrived just then. The meat was freshly cooked, and the kheer had kesar strands sprinkled on top. Fatima had made a lot of effort to please the guests. The rest of the dinner was completed in silence. The militants seemed to enjoy the piping hot food after hiding in the jungles for such a long time. It was a treat, thanks to the generosity of their host, Ahmed Dar, a terrorist sympathizer. It was clear to the militants that this was not something Dar was doing for the first time.

Once they finished their meal, Dar requested to see them off as a mark of respect and walked some distance with them outside the house. As the militants were thanking him and bidding their goodbyes, Dar asked, *'Bura na mane toh aapse ek sawal poochu* (If you don't mind, can I ask you a question)?'

The leader nodded in reply.

Dar said, *'Aap ki tanjeem kya hai? Main bas iss liye pooch raha hu kyunki Azad Kashmir*[5] *se aksar kahi sathi mujhse madad lete hain. Hum toh allah ke bande hai, iss liye dil khol ke madad karte hai. Aapse kuch maang nahin raha, aap to aazadi ke parwane hai, bus itna kehna chahta hoon ki jab bhi aapki "sathiyo" se baat ho, toh humare baare mein zarur bataiyega. Jihad ki raah me adna sahi, sipahi sahi, par jitna ho sakega sathiyo ki madad karunga* (What is your affiliation? I'm just asking because many comrades dedicated to the cause of Azad Kashmir often seek my help. I am just doing Allah's work in setting up the caliphate, that's why I help everyone so generously. I don't wish to ask you for anything. You are freedom fighters, so all I want to tell you is that whenever you talk to your "friends", please tell them about me. I may be small fry in this jihad, but I will help however many friends I can).'

The leader replied without blinking an eye, *'Hum Lashkar se hai. Aap humein yahaan ke Lashkar commander tak pahucha dijiye. Hum aapke ahsanmand rahenge aur aapka paigam zarur border paar bade logo tak pahucha denge* (We are from Lashkar.[6] Please take us to the local Lashkar commander. We will remember your help, and as soon as we cross back over the border, we will definitely tell our bosses about you).'

Dar said, *'Maine aapko pehle hi kaha ki gaon ke nambardar ko kal subah itallah kar dunga, aap kal raat unse issi waqt mil lijiyega—aapka kaam ho jayega. As-Salaam-Alaikum. Khuda hafiz.'* (As I said, I will tell the village headman about you first thing

---

5   The terrorists, militants and enemies of India call Pakistan occupied Kashmir Azad Kashmir.

6   He was referring to Lashkar-e-Taiba.

tomorrow. Then you can meet him during the night, and your work here will be done. Good night and goodbye).'

They hugged each other. Dar had begun walking back towards his house when he suddenly remembered that he didn't even know the name of the man he had invited into his home. He turned back, and called out, 'Bhai jaan, what is your name?'

The leader smiled for the first time and said, 'Tamir! They call me Tamir.'

Once safely inside the jungle, Tamir took out a walkie-talkie, cleverly concealed inside his pitthu, activated it and said, 'Tamir to Nine! Tamir to Nine reporting, sir! Over.'

Bandipora general area
North Kashmir

The following day, early in the morning at around 6 a.m., the nambardar of the village was quietly arrested along with Ahmed Dar, an over-ground worker (OGW)[7] who had been aiding various terrorist organizations. The nambardar had been a local asset, acting as a sleeper cell for Pakistani forces. He had been the one hiring various OGWs, appointing them to carry out multiple activities for various terrorist groups, from providing food and shelter to facilitating the infiltration of foreign terrorists across the Line of Control (LoC) into India. For all of this, he was handsomely rewarded.

The nambardar's arrest was a massive blow to the local terrorist network and their masters across the border, who had suddenly lost contact with their valuable asset.

Meanwhile, Tamir and his men had quietly transformed into their real avatars by changing into their usual Indian Army combat

---

7   Small-time helpers or informants hired by terrorist organizations who run various errands in the particular area.

uniforms. They tied the shoelaces of their combat boots and checked their personal weapons, such as their Tavor and M4 assault rifles. It was moments before the raid commenced. The men loaded ammunition into their secondary weapons before holstering the pistols to their tactical vests. They donned their ballistic bulletproof vests, covered their heads with combat scarves and boarded the Casspir, a mine-resistant ambush-protected vehicle. Their destination was a house in the Bandipora area, believed to be the base of local militants.

With his favourite M-4 assault rifle in his hands, which were clad in half-fingered tactical gloves, Tamir was now properly Capt Tushar Mahajan.[8]

Tushar was a local, a native of Jammu. With his fair skin and tall build, inherited from his Pakistani ancestors, he also looked like he belonged to Kashmir. He was well-accustomed to the local culture and understood the need to mingle with the local population. He could read, write and speak fluent Arabic and Urdu, and was proficient in Kashmiri and Dogri. In 2012, he did a course in the Balti language—a Tibetan tongue spoken by the ethnic Balti people belonging to the Baltistan region of Gilgit–Baltistan in Pakistan. Balti is the second-most-spoken language in that region of Pakistan. Many Balti speakers are also found in Kashmir, Kargil, Karachi, Rawalpindi, Islamabad and Lahore.

Tushar's skills were highly suited for the covert operations of the 9 Para (SF) battalion—the Ghost operators who not only knew how to fire bullets but also how to keep their area safe through their vast networks of spies and informants, as well as through other tactics. Tushar was an excellent covert operative, and his commanding officers exploited this skill well. It was not the first time that the unit had raided a militant base or a hideout in the jungle based on

---

8   Tamir was Capt Mahajan's code name.

Capt Mahajan's inputs—there had been numerous such operations in Jammu and Kashmir in the past as well. Tushar had also been part of various covert operations in the Fire and Fury Corps (Ladakh) area.[9] His skills as an undercover operative ensured the safety of India without even a bullet being fired many a time. His skills also provided safe passage to his teams during ambushes and raids as he walked ahead of the main ambush party, disguised as a local militant, thus managing to keep his teams informed about the situation on the ground.

These were not easy tasks, and the smallest mistake could have cost him his life, but Tushar, which means 'mist', was excellent at evaporating as soon as the operation was over. He was fearless, confident and passionate. Nothing deterred him from serving Mother India like his idol, Shaheed Bhagat Singh.

A true patriot, he bled NINE.

Tushar had always dreamed of joining the Indian Army, ever since he was a boy of six. Hailing from a family of migrants who had left their ancestral homes in Pakistan in the fire of Partition in 1947 and fled to India in their quest of survival, Tushar's passion—to wear the uniform, serve the nation and kill the terrorists who were a threat to the safety of his country and people—was always mixed with some kind of pain.

This is the story of Captain Tushar Mahajan, SC (posthumous), which every citizen should hear with gratitude in their hearts. Youngsters like Tushar are the Bhagat Singhs and Chandrashekhar Azads of our times—and should be treated as such. Tushar's story will

---

9 Tushar was involved in thrilling operations which are still ongoing and are sensitive in nature. They are still confidential and there is no information on them in the public domain. As a result, I have not been able to mention these operations by name or when they occurred—only that Tushar Mahajan risked his life on multiple occasions.

fill your heart with pride as you realize that your freedom comes at a price. It is only the true sons of India, like Capt Tushar Mahajan, who pay that price, sometimes at the cost of their own lives.

~

October 1995
Udhampur
Jammu and Kashmir

'Pabi, please tell me a story,' six-year-old Tushar demanded of his grandmother Jaswanti Devi. She did not deny his request. Tushar was the apple of her eyes, and they spent all their time together. She narrated the only story she knew—the story of their migration from their homeland in Pakistan to India in 1947.

'That year, 1947, was horrible. There were herds and herds of people fleeing. Whole families had been uprooted from their lands. The riots were brutal. There were dead bodies as far as the eye could see. The rivers were no longer blue but red, flowing with blood. Millions of children had been orphaned and women widowed. The shadow of death was everywhere—in the shape of human beings with swords in their hands. Murder was in the air. Chotu, you cannot even imagine the chaos. Nobody deserved this kind of pain. All human beings are entitled to live their lives with peace and dignity.'

Little Tushar listened carefully. The pain in his grandmother's heart pierced his own. He adored his dadi (paternal grandmother), whom everyone lovingly called Pabi. He was young and naive, but his subconscious mind had retained the image of the distraught faces of Kashmiri Pandits who had just migrated from Kashmir.

Tushar was born on 20 April 1989. The constant cries, anguish and mournful faces of the Pandits flooding Jammu was a daily sight in the 1990s. His grandmother's story of his family's migration from Pakistan, and the images of the Kashmiri Pandits, left a lasting impression on him.

Tushar came from a humble background. His mother, Asha, was often busy managing the house, while his father, Dev Raj, was a renowned teacher in Udhampur and the sole breadwinner for their family of five. Dev Raj Gupta's parents had migrated from Pakistan in 1947 and finally settled in Udhampur. Every brick in their house was testament to the struggles and hard work of Dev Raj and his father to provide their family with a roof over their heads, food in their bellies and a name that people uttered with respect.

They did not have much money, but they had the love and respect of the neighbourhood. Dev Raj was a teacher of physics. Today, many doctors and engineers from Udhampur and the nearby areas claim to be old students of 'Dev Raj sir'. Many old students have made the cut and become part of elite institutions as well.

Dev Raj Gupta had great expectations from his sons, Nikhil and Tushar. Both brothers were poles apart—the older brother, Nikhil, was naughtier, while Tushar was quieter, more disciplined and far more academically brilliant. Dev Raj dreamed that one day he would join the Indian Administrative Services, never suspecting that Tushar had different ideas and aspirations altogether. He would engage in mock sword-fights and gunfights with his best friend, Shwetaketu Singh Jamwal. The two friends would listen to stories of Maharana Pratap, Prithviraj Chauhan and Rani Lakshmibai from Shwetaketu's grandfather, who had once served in the Indian Army. The two boys lived close to each other, studied in the same school, sat on the same bench and shared many interests and hobbies.

When I met Shwetaketu Singh Jamwal to record Capt Mahajan's story, he showed me what true friendships look like. Shwetaketu had known Tushar since they were both in class two. The tragedy is that instead of dancing at Tushar's wedding, Shwetaketu had to carry the diya before his best friend's casket during his last rites instead. It had been brutal to pick up the ashes from his best friend's pyre. But this did not deter Shwetaketu from the path he chose in the memory of his best friend. While he practises as an advocate in the Jammu wing of High Court of Jammu & Kashmir and Ladakh, Shwetaketu has stood firm with Uncle Devraj and Aunty Asha during their toughest times.

Shwetaketu told me, 'Swapnil, I have been raised hearing stories of Bhagat Singh, but I never knew I would have the privilege of being raised alongside a legend just like him. Now I realize how different Tushar was from the beginning. Patriotism was in his blood. He hated terrorists even at the age of six. Adventure and bravery were second nature to him. While he was never involved in street fights, unlike most boys of that age, his first instinct was always to protect others. The Kargil War greatly impacted his mind, and Capt Saurabh Kalia inspired him. We always discussed how he would have endured torture in the nation's service. He never missed an episode of *Swaraj*, the TV show on Doordarshan which featured Indian freedom fighters. Veer Bhagat Singh was his ultimate icon. He was never fascinated by Bollywood actors or sports stars even in his childhood.'

Shwetaketu paused for a minute, his face lit up with nostalgia and pride, and said, 'At that tender age, while growing up, we would discuss freedom fighters, wars and the people who gave so much for their nation. But we never guessed that one day, Tushar himself would join that league and make the supreme sacrifice—leaving his parents, friends and dreams behind in the service of the country he loved so much.'

Perhaps it was written in the stars. Perhaps Tushar had known all along. Perhaps that was why, when he was in class two, when his English teacher asked them all to write an essay on the topic 'My Aim in Life', Tushar wrote a beautiful essay on how he wanted to become an Army officer and serve his nation. Many other students, including Shwetaketu, wrote the same, but as they grew older, their aspirations changed. Only Tushar's remained the same. Until class twelve, every time he was asked to write an essay or participate in debates about his life's goals, he would persistently talk about wearing the olive-green uniform of the Indian Army.

His father, Dev Raj Gupta, smiled as he talked to me and said, 'Beta, Tushar was always determined. If he said he wanted to do something, he would do it. He was highly disciplined and did not like wasting time. He would fold his school uniform, polish his shoes, keep away all his books neatly and even switch off the lights to save electricity. We joked that he behaved like an Army officer without even realizing that one day he would become a rather exceptional one.'

Tushar was academically brilliant and always topped his examinations. He was also the cricket captain and the head boy of his school. Many girls liked him, but he never paid much heed to them. Shwetaketu laughed as he told me how he had once, in class eleven, pushed Tushar to get himself a girlfriend only because Shwetketu also had a girlfriend. Tushar eventually did date a girl in school, but whenever they went out, he would yawn or sleep or concentrate on eating. Eventually, the girl dumped him out of boredom, while Tushar, of course, was relieved.

Nobody knew that he applied for the Services Selection Board[10] immediately after finishing class twelve. He never went to any coaching classes or prepared for any of it. Dev Raj and Asha Gupta were returning from Pune, where Nikhil was studying engineering, when Tushar called them to give them the good news. The couple was baffled. Nobody from the family had ever joined the armed forces. But Tushar was adamant.

---

10 The Services Selection Board (SSB) is an organization that assesses candidates for becoming officers in the Indian Armed Forces.

In June 2006, Tushar joined one of the finest military institutions in the world: the National Defence Academy (NDA) in Pune, Maharashtra.

~

June 2006
NDA
Khadakwasla, Pune
116-NDA Course

Tushar was in the first battalion: the Alpha squadron, his home for the next three years. The quadron is more like a temple for cadets in the NDA. The cadets go to any extent to prove themselves by defending the squadron colours, perhaps the first step to imbibing the esprit de corps so necessary for Army life. Tushar's journey to become a true officer had now started, but the path was arduous.

Training meant that the seventeen-year-old boy had to start his day at 4 a.m. Many a time, he was not able to even sleep. It was the norm at the NDA that a combination of the training curriculum and the entire establishment broke each boy in order to turn him into a man who could be entrusted with the nation's safety. Rigorous and harsh treatment was required to turn the boys into solid leaders and soldiers.

As a result, Tushar's day would start with PT, drills and academic classes followed by various punishments by ustads and seniors rest of the day. Sometimes he would undergo a 'water session' early in the morning, where he would have to drink water to the point where he puked, only to be forced to drink more water and puke again. On other days, he would undergo a 'heat session', which meant rolling on the concrete roads of Khadkwasla Pune under the hot sun until his back bled profusely or burned in the heat. There was no dearth of punishments during those three years in the Academy—side roll, cream roll, maharaja, whiskey, helicopter.[11]

At the time, Tushar was not very physically strong, but he was disciplined and extremely hard-working. Eventually, he started excelling in every category. He was an outstanding cross-country runner and was also good at boxing. His skills helped his company win many competitions. Soon, he was appointed to the post of Divisional Cadet Captain of his squadron, a prestigious appointment bestowed only upon the crème de la crème in the squadron.

Since childhood, Tushar had been a determined boy, but these three years in the NDA had turned him into a man who knew how to achieve his objectives, perceive things differently, analyse a situation and come up with a solution. This self-actualization of his inner strength and weaknesses helped him decide the goals he looked forward to achieving in the future.

In June 2009, Tushar's parents proudly attended his passing-out parade. It was only that night—when they attended the official dinner in the historic officers' mess at NDA Khadakwasla and dined with the other cadets and their parents—that it really dawned on them that their son was now an officer in the Indian Army. That evening, as they ate the lavish seven-course English dinner and listened to the band playing melodiously while their son sat next to them, elegant and comfortable in his uniform, their hearts swelled with pride.

In July 2009, Tushar joined the Indian Military Academy (IMA) in Dehradun, Uttarakhand. He was assigned to Alamein Company, named after the Battle of Alamein. The companies in the IMA are akin to squadrons in the NDA. But after the rigours of the NDA, the IMA seemed like a cakewalk. He was more relaxed and confident

---

11 These punishments are part of the military curriculum and are meant to toughen up young boys and turn them into true soldiers ready to face extreme situations, both mentally and physically.

this time around and made many friends. Training at the IMA lasts for a year, and right before the passing-out parade, the gentlemen cadets are asked to fill in their choice of arms. It came as a shock to everyone, including Tushar's parents (who had insistently asked him to opt for support arms like ordinance or ASC), when he volunteered for 9 Para (SF).

Only Shwetaketu had some inkling of Tushar's plans. He said, 'When Tushar visited me during his IMA term break, he had this poster with the words "Welcome to 90 days of hell" written on it, with a deadly looking skull wearing a maroon beret and a dagger in its mouth. I had no idea that this was the logo of 9 Para (SF). I was curious and asked him about it. That's when he told me that he wanted to do something big with his life. He said that the Special Forces were the most elite of forces from around the world. He looked confident and said he would be joining that league soon.'

Tushar was from Jammu and Kashmir, the state nearest to the LoC, where the base headquarters of the 9 Para (SF) battalion[12] is located. He had seen the (SF) operatives at work since his childhood and was fascinated by their legend. More importantly, he had witnessed the suffering of the Kashmiri Pandits in Jammu. He had felt his Dadi's pain. His decision of joining the 9 Para (SF) was the culmination of his experiences since childhood.

While I was researching Capt Tushar Mahajan's life, I realized that even though Tushar had been snatched away from his loved ones in a brutal twist of fate, he had been a highly motivated, meticulous person who always planned things in advance. Whatever he chose to do, he did it with the utmost conviction—right from his schooldays, where he stuck to just one goal, that of becoming an Army officer,

12 https://testbook.com/defence/9th-para-sf (where is 9 PARA SF located); https://www.thedefencearchive.com/post/9-para-sf (history and origin of 9 Para SF)

to his Academy days, where he always performed exceptionally, or even his probation, where he was one among twelve selected out of a hundred candidates. Tushar did not believe in the ordinary.

~

July 2010
9 Para (SF)
Probation period
Jammu and Kashmir

The 9 Para (SF) battalion specializes in mountain warfare. Since they are permanently deployed close to the LoC and also operates in the LAC, this battalion is also always conducting active operations and is involved in regular encounters. The selection criteria for the unit are different from most other SF battalions. The 9 Para (SF) officers' trainees are put to test under more difficult circumstances than other jawans because, eventually, they will have to lead the troops, the team and the entire battalion. As a result, future stalwarts act as guiding lights for new generations of 9 Para operatives. There are many things at stake—from regimental bonds to the safety and sovereignty of the nation. This raises the attrition rate of the unit to close to 99.9 per cent.

Lt Tushar Mahajan was one of the two officers and ten jawans to join the 9 Para (SF) in 2010 out of 100 probationers.

Immediately after the IMA's passing-out parade, Tushar returned home. Along with him, four other cadets who were now lieutenents had also opted for the 9 Para (SF), and they were banking on Tushar for further guidance. As the local boy, Tushar had invited them to his place a day before the probation period began. The five boys, all of them in their early twenties, enjoyed the last of their 'happy' days at Tushar's home before they began their quest to become part of a legendary battalion. Asha Gupta was overjoyed at the prospect of looking after the boys and fed them heartily with aloo paranthas, some famous Jammu kaladi and rajma.

The following day, Dev Raj Gupta dropped the boys off cheerfully at the gates of the 9 Para battalion base. Only then did it dawn on them that they were going to be on their own for the first time. In total, there were about a hundred boys, including both officers and jawans. On the very first day, they were allotted kits and individual AK-47 rifles, which they were supposed to keep with them at all times. The rifle had to be kept clean and shining. One scratch would mean harsh punishments. There was no distinction between day and night because the training levels had increased multifold from what they faced in academies. They would be ordered to run, train at any time and under any circumstance. Each moment, they were tested physically, psychologically and academically, with assignment after assignment.

Tushar's fellow probationers remember him with great fondness. Naik Shamshad shared with great nostalgia, '*Saab mein baat toh thi, kuch hatke the, tabhi toh NINE mein select hue* (There was something about Saab; that's why he was selected to the NINE).'

Naik Shamshad continued, 'What I liked most about him was that he was kind-hearted and caring. During probation, he would help the jawans understand the theoretical classes. I also remember how he tended to me during one of the speed marches when my ankle was broken. There was a time limit for everything. You either fail or pass. Every day, boys were sent home. But he stopped and chose to help me out rather than continuing on his speed march.'

This wasn't the first time that Tushar went out of his way to help someone. During my research, I uncovered many other touching incidents of Tushar's generosity. His academy mates, fellow probationers, civilian friends, fellow officers and the chaps who served in his command all vouch for his selflessness and kindness.

The speed marches, long route marches, swimming and other physical exercises were not just brutal, they were the toughest in the Indian Army. The Battle Physical Efficiency Test and physical training were conducted at impossibly high standards, and every week, the boys would be taught a new skill. Tushar acquired navigational skills, when he learned all about how to use GPS, a compass, and other sophisticated devices; weaponry skills, when he mastered the use of personal and support weapons; and honed his firing skills on weapons like PKMGs (general-purpose machine guns), under-barrel grenade launchers, Tavor MP4 rocket launchers, and more. He also became an expert at providing medical aid to a dying soldier, the dos and don'ts of saving a life in minimal time, and even evacuating casualties.

The outdoor exercises, considered sheer torture, were specifically designed to test the mental capabilities of the boys. They would be sent alone into harsh jungles for several days with complete combat load and ammunition but without any food or ration. The nights were dreaded and the days were cruel. Many probationers broke down and radioed back to be evacuated, giving up on their dream of joining the league of headhunters.

Then came stress week, where the rest of the probationers were kept in isolation in the dark, without food, and often beaten. This was done to check their behavioural patterns. The battalion needed men with remarkable mental resilience. Many could not survive stress week and were immediately sent back.

And then finally came the most dreaded mountain-running exercise, where probationers were required to run with a full battle load to complete a long route march on a steep mountain within a limited time. The unique run was so backbreaking that only a few men could complete it. But that is what the 9 Para (SF) is all about—conducting operations at high altitudes. A mistake in choosing probationers could lead to severe repercussions for the whole unit.

Almost every day, some of the boys would be sent back. Only the best of the best had the right to wear the maroon beret and adorn that Balidan badge on his chest. The NINE were a league apart. Tushar kept surviving week after week, steadily.

At the end of ninety days, only a few probationers were left. Twelve men—two officers and ten jawans. They were selected to be an integral part of the league of the ghosts of the valley and were bedecked in maroon in a happy ceremony. Beer flowed and the push-ups never stopped.

That was the day that Tushar called his parents for the first time in ninety days. He also called the love of his life—Seema.[13]

~

December 2010
Jammu and Kashmir

'Hello, light of my life!' exclaimed Tushar. He had called Seema as soon as he could after his marooning.

'Oh my God, Tush! I was counting the days to your call! It has been months since you last called. Where have you been?' Seema asked.

Seema and Tushar had known each other since Tushar's NDA days. She was from Jammu, and they had met at a common friend's birthday party. Casual interactions had soon blossomed into something more, even though Tushar could not manage much time for romance with his hectic Academy schedule. But he and Seema talked for long hours—sometimes the entire night—on the phone.

By the end of his IMA training, their relationship was clear—they were committed to each other. But since Tushar had to immediately start his probation at 9 Para (SF), he could not always stay in touch

---

13 Name changed to protect privacy

with Seema, often leaving her clueless about his whereabouts. But he would call her whenever possible, and they would always make plans to meet.

Seema remembered, 'I was worried to see him so thin. When we met for the first time after his probation, his shaved head and weight loss scared me. But he made me understand how completing probation in the 9 Para was a huge achievement in itself. Though looks never mattered to me, his typical fauji sophistication always took centre stage. Never once did he meet me empty-handed. He always carried precious little things like chocolates, flowers or even teddy bears. We hardly ever met, though, even when he was posted in Jammu, because of his challenging deployments at the forward posts and operations. But he ensured that he made me feel special with all those small gestures whenever he met me. He would pull out chairs for me and treat me like a princess. I would be on cloud nine those days.'

There would be times when Tushar's phone would be switched off for weeks, but it never stopped Seema from loving him. Instead, she led a simple life—home to office and back was an average day. Tushar made sure to call her for even for two minutes every time he had to leave for an operation where he would be out of touch.

However, Seema confessed to me that she would be scared during these times and would pray for his safe return. She developed a keen interest in the SF and their modes of operation in those days, and she googled, read and researched as much as she could about them. Unfortunately, though, all this only served to increase her anxiety. She would cry incessantly whenever Tushar returned after weeks in the jungle, with Tushar consoling her and asking her to be brave like his mother, Asha. After all, she was a Special Forces operative's girlfriend. He needed her to be strong, he said.

Seema tried her best. She kept herself busy in his constant absence but never once did the thought of breaking up with him cross her mind—even at the tender age of twenty-one. The bond they shared was very special.

In 2011, Tushar completed his basic parachute-jumping course from Agra, his Young Officers course from Mhow and the mandatory commando course at the Infantry School in Belgaum. Every degree he earned and every course he completed honed his leadership and operational skills. By the end of the year, he had rejoined his unit and been deployed to Ladakh.

~

June 2012
Somewhere around the Line of Actual Control (LAC)
Ladakh

Capt Tushar Mahajan was in Ladakh, somewhere close to the LAC. It was a high-altitude posting, almost at the edge of the Indian border, where the base altitude itself was about 10,000 feet and the environment was harsh and rarified. Tushar's team carried out many training and reconnaissance activities, focusing on creating the best SF operatives who could operate swiftly at such altitudes, even at short notice. New war tactics, involving the testing of equipment and weapons, were also being tried out.

Tushar and his men would climb to great heights at sub-zero temperatures over the breathtaking Ladakh mountains. The clear blue skies and the pristine environment in the cold desert was heavenly, but the region's fragile topography presented many challenges. Tushar needed to be mentally and physically fit to uplift the troop's morale, as the harsh weather took a toll on the men, not just physically but also psychologically.

Tushar always ensured that he met with the troops in his capacity as Troop Commander as soon as he returned to the base after his missions. His colleagues told me that this was special because, usually, after completing an operation in such an environment, the immediate physical need of the body is to rest or sleep. But Tushar was a true leader, and his kindness was on display at every point.

During one such surveillance and recce operation, Tushar and his men were climbing a steep mountain at high altitudes—a place where even the slightest mistake could result in loss of life. For many seasoned mountaineers, summitting this kind of a peak was considered a victory. But for Tushar and his team, this was but an ordinary task as they regularly climbed to such heights for their various missions. However, this time around, the area in question was completely unexplored—nobody had gone there before. They had no frame of reference to attempt this mission, but Tushar wanted to check the area anyway and mark it so that, if and when the time came, his inputs could help prepare the blueprint for future attacks on the enemy. He also wanted to check the performance of some new surveillance equipment at that altitude.

As they made their way up the peak, some of the boys started to feel ill—they were suffering from altitude sickness, which weakens your mind and body, leaving you at the mercy of the mountains. Usually, the only way out of such circumstances is leaving the afflicted men behind and calling for a rescue party to help them—heptrs[14] are usually out of the question because they cannot operate in these covert areas so close to the enemy's base. But, for Tushar, his men mattered the most.

He quickly radioed his Team Commander, saying, 'Delta to base, Delta to base. Over!'

His Team Commander came on the line and replied, 'Base to Delta, base to Delta. Report!'

---

14 Army helicopters

Tushar replied, 'Sir! We are at the peak of the assigned target. We successfully climbed it, but some of the boys fell sick. It will be nightfall soon, and the temperature will plummet by several degrees. My men cannot climb down, and they won't survive the night. Request directions.'

There was a brief pause on the line before the Team Commander said, 'Even if I send a rescue party, I can only do it at first light tomorrow. Helicopters cannot fly there, nor would they be able to locate you in those mountains. I am afraid there is not much I can do to assist you. You are the best person to decide what to do next. I feel it would be good to climb down as quickly as possible with the rest of your men and leave the sick men in some shelter. Ask them to wait until the rescue party arrives. That is the best advice I can give you. Over and out!'

It was now a life-or-death decision for twenty-three-year-old Capt Tushar Mahajan. They were at an altitude where there was no way ahead nor any way back down that could be used for evacuation. Recceing the area carefully, Tushar discovered a cliff that could be used to evacuate the men. Along with a few of his men he lifted and evacuated their sick comrades one by one. Eventually, Tushar managed to reach a safe area, which they could use as a harbour for the night. And the following day, he successfully evacuated his troops without any casualties. It was a Herculean—almost impossible—task. But he did it.

It was a tough posting for Capt Tushar Mahajan, especially because it was difficult to connect with his loved ones. Communication towers were a rarity. There was one BSNL tower at the base where he would go and wait desperately to get a signal whenever he got some free time. The signal strength would always displayed maximum connectivity, yet calls wouldn't go through. Sometimes, messages took days to be delivered. Tushar began talking anxiously with his colleagues about how he worried that Seema would dump him.

It is a fact that Seema was indeed unhappy. The two had hardly met before, and now, even talking on the phone was a rare luxury. As a civilian, she tried hard to understand Tushar's situation, but the passing of days, weeks and, sometimes, months began taking a toll on her. She would think hard about ending things, but just one brief phone call from Tushar was enough to weaken her resolve again. She had fallen in love with a man who could not—out of duty—make her his top priority.

Still, Seema persisted—her heart refused to heed what her head was saying. Someone once said that only a woman of steel can love a man in uniform—they are not wrong. These remarkable women, at the risk of continuing to hurt themselves, keep hope and love alive in the toughest of circumstances. It is these women who keep the men at the borders going. These are the women to whom these men look forward to return to one day. They inspire the men at the border to give their best even during the toughest of circumstances.

As the famous English writer and philosopher Gilbert Keith Chesterton had once said, 'The true soldier fights not because he hates what is in front of him, but because he loves what is behind him.'

With the passage of time, even this difficult period eased. After a year, the tough Ladakh posting was over. Tushar and his team were ordered to fall back to the Jammu and Kashmir area.

~

June 2013
Lolab Valley
Jammu and Kashmir

Op Shelter was a fortified base, constructed of prefabricated sheets over a cemented floor in the middle of the jungle in Lolab Valley. These fortified bases, where even a bird could not enter because of the high security, acted as a home to the forces deployed in the Indian wilderness. One of these temporary accommodations was occupied by Capt Tushar Mahajan and his roommate, who was a senior SF officer from the same unit.

Tushar had managed to convert his basic room into a modern space. There were LED lights hanging on the walls, and the little wooden table in the corner was laden with neat stacks of books. Tushar had been an avid reader since childhood, and always carried piles of books with him wherever he went. He had recently bought a Kindle on which he stored hundreds of books on philosophy, religion, spirituality, wars and history. Right above his single bed were wooden stands with hooks from which hung all his weapons, including his favourite M4 assault rifle. Tushar had a peculiar preference when it came to decor, considering they were in an operation shelter. I asked his roommate about it.

He laughed and said, 'Ma'am, Tushar was certainly different. He was a fiercely patriotic person. He was a true nationalist. His weapons were always ready. He deeply idolized Sardar Bhagat Singh—his pictures were hung up all over his room along with the tiranga. Books were his other companions. After he immortalized himself, we opened up his trunk and found it to be full of books—from Swami Vivekananda and Bhagat Singh to Anne Frank and Leo Tolstoy. If you look over the kind of books he liked to read, you can see what kind of a man he was, why he was propelled to achieve the impossible for his nation and its people. We never realized it when he was with us because, then, the other side of Tushar would come out. He loved listening to English songs at full volume and dancing like crazy. He could dance anywhere, even in the middle of the road, with the car stereo blaring at full blast. I remember cooking chicken with him, out in the open. We would put onions into Haldiram's aloo bhujia and eat it with Old Monk, often when in an op shelter. We would always have music playing in the background while we talked about life and our goals. Now it seems surreal that I shared some time and space with a legend like Tushar.'

It is tragic irony that young Tushar would eventually return home wrapped in the same 6-feet long and 3-feet wide tiranga that he had put up in his room from Delhi. Today, the flag is framed and hangs in his house at Udhampur. His room is a museum of his things, maintained with love and pride by his parents. When I researched his story, I wandered through that house for a day—by the time I left, it felt surreal, mingled with pride. You can feel his presence in his room.

Tushar was actively involved in various surveillance and reconnaissance operations in the Valley—sometimes he was a covert operative, while other times he simply hid in the jungle for months, observing a specific target area or even a potential threat. He would lay tactically brilliant ambushes, camouflaging himself for days along with his squad, awaiting the arrival of the militants. There were instances when he carried out high-risk surveillance and recce operations near the LoC, where he would sit for days and observe enemy posts, find routes for quick raids if the need arose and also locate possible infiltrating routes or areas to lay mines. He also carried out surveillance of various terror launch-pads in PoK for several months, gathering valuable information about terrorist activities, the terrain, and the demography and activities of the local Pakistanis. All these operations were conducted in anticipation of any surgical strikes or other potential enemy threats in the future. Operational preparedness is a continuous activity, and the Indian Army believes the more you sweat in peace, the less you bleed in war. It was a never-ending job, and Tushar's valuable inputs helped his unit impeccably plan many operations.

This level of care and attention to detail was what made Tushar who he was. His childhood friend Shwetaketu Singh Jamwal told me how, during the few day trips they had taken together, Tushar would make sure to change the car's number plate several times.

Upon being asked why he was doing so, he would cautiously explain that he might be on the radar of the enemy nation, so he needed to cover his tracks. Many of his relatives affirmed how he never left a digital footprint while shopping. Even Seema, his girlfriend, told me how he would suddenly appear out of the blue outside her office to meet her—dishevelled and scary, in a long beard and a skullcap. Everyone at work knew about Seema's fauji boyfriend. They had assumed he would be charming and sophisticated—so to see her with a man who looked more like a terrorist than anything else was shocking. To add to Seema's misery, she could not tell anyone why her boyfriend looked the way he did. In fact, she told me that for a long time, people actually suspected her of seeing someone else, or possibly even acting as a honeytrap.

Alas, for security reasons, there is no proof and nor will there ever be proof. But there is no doubt that Capt Tushar Mahajan was an excellent covert operative who was part of hundreds of covert operations. In these crucial cross-border missions, he played various roles and held different identities. My sources for his story have established how deep and valuable an asset Mahajan was, in various capacities. Still, being an asset was only one part of his multidimensional personality. Another aspect was that he was a saviour too.

~

March 2014
Chenab River, Ramban area
Kishtwar
Jammu and Kashmir

Tushar was an excellent combat diver. You can still find pictures of him standing in uniform, with the combat-diving badge pinned proudly to his chest—one of the most coveted badges in the armed forces. Combat divers are some of the most robust and fit soldiers in the forces, with the ability, skill and passion to carry out any underwater attack. Small wonder, then, that the attrition rate in the combat-underwater diving course conducted on the *INS Kochi* is almost 99.9 per cent. Only a select few make it—those few are the best of the best.

Capt Mahajan was one such combat diver who made it through the arduous three-month course at INS Venduruthy in Kochi. As a diver and a leader, he was part of many underwater rescue operations in the Jammu and Kashmir region.

In March 2014, a civilian truck suffered an accident and sank in the feisty Chenab River, near the Ramban area. There was a massive uproar. Hundreds of people gathered on the riverbank to receive the bodies of the truck driver and the two other people who had drowned along with the truck. The civil administration was finding it challenging to locate the bodies. They had neither the equipment nor the expertise to carry out the risky diving mission into the dreaded Chenab, which was notorious for its strong currents. They requested the nearby 9 Para (SF) unit to carry out the task.

Capt Tushar Mahajan, along with his squad, reached the Chenab River clad in special neoprene diving suits to protect them from contracting hypothermia from the icy dark river water. Customized underwater diving gear and heavy, twin-air oxygen cylinders were strapped onto their bodies, along with underwater searchlights, compasses, depth gauges and other sophisticated diving equipment. To swim in the water carrying that much weight would be an impossible task for any normal human being—but these were combat underwater divers of the Indian Army, specifically selected and trained to deal with unbelievable situations.

The first challenge for Tushar was to find the exact spot where the truck had sunk. He knew the vehicle would have disintegrated due to the strong currents of the Chenab. Tushar, along with some of his men, eventually found the target location at the bottom of the river and, as suspected, found it filled with sharp iron rods and giant boulders.

It was pitch-dark under water, and the men could feel the excruciating stab of the piercing river current even through their neoprene suits. With the help of an underwater searchlight, they kept searching for bodies in the area. On the first day, they could not locate any bodies despite covering an area of several kilometres in the river. This went on for two more days and, eventually, the team managed to locate a body on the third day.

While the other bodies had drifted away to impossible distances and unknown locations, they were able to recover the body of the truck driver because it had got stuck among rocks. However, to pull the cadaver out of the river in conditions of zero visibility and with the steep rocky terrain around, was a huge challenge. One mistake, and the diver could also lose his life. It takes the highest level of expertise to pull off this stunt.

Tushar took it on himself to extricate the body and bring it back to the surface safely. He tied the dead body to his own and, in an exceptional display of skill, eventually pulled out the body.

The crowd outside suddenly turned emotional. Relatives were crying out loud—but they were cries of relief, because now they could perform the last rites of their loved one. Everyone applauded the might of the brave diving team. Some locals even touched their feet as a sign of respect, making Tushar uncomfortable. This was not the first time that he had performed an underwater rescue operation. As soon as he could, he vanished from the scene, leaving the other forces and civil administration to bask in the limelight.

~

21 February 2016
Pampore
Jammu and Kashmir

At around 2.30 a.m. on 21 February 2016, the 9 Para (SF) received a request to send a team to lead a building intervention operation at the Jammu and Kashmir Entrepreneurship Development Institute (JKEDI) in Pampore. Several casualties had been reported in the ongoing operation, which had begun on 20 February 2016, when an unknown number of terrorists had attacked a convoy of Central Reserve Police Force (CRPF) personnel. There had been two casualties and several injured on the CRPF side during the ambush. As soon as the CRPF personnel retaliated against the terrorists, they had fled to the nearby JKEDI building. The enemy attack was so well planned tactically that the security forces took some time to grasp that the encounter would go on for a longer period of time. The terrorists had by then vacated the open spaces where it had been easier to kill them. The nearby Rashtriya Rifles team, which had reached the target area along with the CRPF officials, realized that they were looking at a potential hostage situation as the JKEDI building reportedly had hundreds of Kashmiri locals working inside. It seemed like a dead end.

Not only are building intervention operations one of the most challenging military operations to deal with, now there was also a hostage crisis to tackle. Room or building operations mean that even a tiny number of terrorists can potentially cause a considerable number of casualties. There is no surprise element and no vantage point for security personnel, since there is only one entry point into the room. All it takes is just one bullet from the terrorists hiding under a bed or a sofa, or even behind the curtains, to find its mark. The terrorists are on a suicide mission—they want to create maximum havoc before they are killed or captured.[15]

The hostage situation was a huge deal for the terrorists—it would lead to a lot of media coverage and more publicity. On the other hand, the situation meant crippling pressure on the forces, coupled with a limited area of operation.

These were no ordinary terrorists—they were highly indoctrinated and trained. They had clearly planned meticulously before attacking the CRPF convoy. Later, it was also revealed that these three Lashkar-e-Taiba terrorists had stayed in Pampore for quite some time. They had recced the building and, as a result, knew its layout and had planned all their vantage points before the actual firefight.

By then, reports had started trickling in that Capt Pawan Kumar from 10 Para (SF) had attained veergati. Initially, the unit, located near the target area, had been called after the CRPF and the Rashtriya Rifles had lost control of the situation. Many attempts to enter the building had failed and several security personnel had been injured in their quest to enter and rescue the civilians trapped inside. The 10 Para (SF) team immediately rushed to the scene, but once the brave Capt Pawan Kumar attained veergati, the operation was transferred to the 9 Para (SF).

Once the Commanding Officer of the 9 Para (SF) received the request and analysed the ground reports, he knew he had to choose the best man for this risky operation. He immediately called Capt Tushar Mahajan, who was near the area, and his team. The

---

15 A similar type of building intervention exercise happened during 26/11 Mumbai attacks, when a small group of terrorists occupied a building and took advantage of the hostage situation. For similar reasons of protocol of building intervention exercises, it took some time to flush them out. It is said that casualties on the side of security forces are always a strong possibility during such operations.

Commanding Officer briefed him on the situation and asked him to lead, plan and execute the building intervention operation to destroy the terrorists.

Capt Tushar Mahajan had attended an Indo–US joint exercise in 2014, and had learned building and room intervention tactics from the US Navy SEAL special forces. He had been the only one from 9 Para to attend the exercise that year, and he had learned a lot about various hostage-rescue operations and building clearance and room intervention techniques. He knew how to sanitize the building and clear it from room to room, which meant eliminating terrorists, removing any threat, etc. He knew about tackling hostage situations and how the team could enter the building. He was also well versed with the different types of equipment and weapons to be used during such urban-warfare scenarios.

Tushar had prepared a field manual crucial to devising in-room intervention techniques, which was eventually made part of the battalion's training routine and curriculum. The unit also used his instructions in live operations. He was an authority in 9 Para (SF) when it came to building intervention exercises—and, as fate would have it, he ironically sacrificed his life in one such operation too.

Tushar was happy to be tasked with leading this exciting operation—he and his squad were supposed to move to Leh in May 2016, where there would be no chances of terrorist encounters and only more dull surveillance and reconnaissance operations.

In the middle of preparing and leaving for the operation, Tushar took the time to call his Team Commander, who was on leave in Delhi to celebrate his wedding anniversary. A sleepy and irritated Sana Sheikh[16] picked up the phone at 4 a.m.

---

16 Name changed to protect privacy.

Tushar chirped, 'What is this, ma'am? Who sleeps on their anniversary? Where is Sir? Did you guys not go to a party? You're in Delhi, make use of it.'

Sana pretended to be angry, even though she was helpless against his boyish charm. 'What is this, Tushar? Is this the time to call? You could have wished in the morning.'

Tushar replied, 'No, ma'am! Kahaan (How would I)? I am on my way to an operation. *Tagda contact hone wala hai* (I am expecting a big contact soon). But you forget all these things and focus on the cake you will bake for me when I come to your place. I will catch up with you and Sir before leaving for Leh. You are not cooking anything nor sending anything for us these days.'

Sana laughed and promised to cook all the delicacies Tushar desired and handed the phone to her husband. Tushar quickly briefed the officer about the impending contact, exchanged notes about the operation and bid him goodbye.

Neither of the men guessed that within hours, there would be another call. The officer would have to rush to Pampore the following day itself, taking over an operation Tushar was never destined to complete.

By 9 a.m., the teams were in place at Pampore. The Commanding Officer had also reached the scene via helicopter. The Team Commander in Delhi had received a call from the CO by that time and had left immediately on the next available flight to Kashmir. He landed at Srinagar and was immediately received by his men, who took him to the nearest secret warehouse of the 9 Para (SF), where he equipped himself and left on a Tata Safari for Pampore.

Meanwhile, Tushar realized that the locals outside were providing information to the terrorists inside the building. There was no other way to explain the clean shots they were taking. His other concern was the media. He realized that the terrorists might have been watching television news channels in order to glean additional information (as well as for the adrenaline boost from watching their operations being telecast live).

One of Tushar's comrades, who had participated in the operation, told me, 'Pampore is just fifteen kilometres away from Srinagar and is notorious for antinational activities. Not only were locals pelting stones and raising slogans, the unwanted media coverage was also troubling. All this puts a lot of pressure on us and provides publicity to the terrorists who then die glorious deaths, in their eyes. Anonymity and no publicity for their operations are their biggest fear. Media coverage is a dream come true. They were firing so accurately at us that we understood that they were getting live inputs on our movements. Tushar realized this quickly and we were able to request mobile networks to shut down (an uncommon occurrence at the time) and stop the media from telecasting the operations live.'

Tushar carried out a reconnaissance of the building and laid out an excellent plan to cordon off the building and execute the mission. His first act was to place snipers all around, targeting the windows, so that the assault team could enter the building without any injury. This is what had happened in the past—every time the forces had tried to enter a building, they were faced with a volley of bullets. Tushar then used grenade launchers to make holes just beside the windows, so that the snipers had proper visibility for an accurate shot inside the building. He then briefed the team to make their entry from both sides of the establishment to confuse the terrorists about their exact movements. Finally, he asked his squad to load charges on the doors so they could be blasted down. Needless to say, this was tactically a brilliant plan.

It was during these moments of planning that Tushar's Team Commander reached the target area with whom he had a word last night. Tushar heaved a sigh of relief and said, 'Accha hua, sir, aap aa gaye. Confidence nahi aa raha hai. Please mere side hi rehna (Sir, it's good that you've come. I'm not feeling confident about this. Please stay by my side).'

This was a shocking statement. Normally, only soldiers who are confident about the mission are made a part of it, for obvious reasons. But this was Tushar, the expert in room intervention techniques. Before the Team Commander could reply, he received a call from the CO at the outer cordon, who was handling the administrative issues and other logistics required for the operation. By then, 9 Para (SF) had also arranged an impressive array of weapons needed for the operation. Ideally, there was nothing that could go wrong.

But sometimes, life plays cruel jokes. Did Tushar have a premonition about what was coming his way in just a few hours?

Many of his teammates and friends believe that he did. Shwetaketu told me how Tushar had come home on leave in December 2015 and told him that there had been too much infiltration from across the border recently, and that they were expecting heavy casualties in 2016. They had no inkling then that Tushar's name would also be on that list.

In that same period of leave, Tushar had also promised Seema an official engagement in April 2016, and a wedding soon after. Their relationship had seen many ups and downs and even break-ups. But every time, they had returned to each other. Tushar's family had initially disapproved of the match because Nikhil, his elder brother, was not yet married. Seema's family, on the other hand, was under constant pressure to marry her off soon. Eventually, though, Tushar and Seema had overcome their problems, and both their families had agreed to their alliance.

No wonder, then, that Seema was so elated the last time they had met, when Tushar had promised to come back to her soon. Their roka[17] was scheduled for April 2016, and just thinking about the

---

17 Engagement ceremony.

festivities ahead—the ceremony, the shopping and the honeymoon—
led to butterflies in Seema's stomach. Sadly, sometimes, love, devotion
and loyalty aren't enough.

Back in Pampore, the doors were blasted down. The terrorists
attempted to fire at the SF operatives, but well-placed snipers provided
the team with excellent cover. The terrorists realized something had
changed, so they immediately rushed to vantage points.

The two teams stormed the building while the snipers opened
fire to cover them. Using their highly specialized combat skills, the
teams began securing the building, floor by floor and room by room.
The JKEDI building was a plush, modern construction, sprawling
over 10,000 sq. feet. It was four floors with hundreds of classrooms
and offices, several halls, storerooms and washrooms. The roof
was sloping, and the front of the building was all glass. There was
a massive stairwell which ran the length of the building from the
bottom to the top, allowing people at the top to take an easy peek
down at everything happening below—a huge issue during a tactical
operation.

Terrorists in the well-covered stairwell could easily target the
forces on the ground floors and quickly change their positions too.
An additional impediment was that the bullets being fired by the
forces also would not be able to easily reach them.

One of the officers, an expert in building intervention tactics,
told me, 'Ma'am, no matter how many types of equipment or drones
we have, until and unless we intervene in every room on every floor
physically, the operation is never over. Our chaps know that out of
these hundreds of random rooms, there will always be one room
where well-hidden and well-positioned terrorists will have the first
opportunity to fire on us. Many times, we will be exposed. After all,
the terrorists know that we are coming to find and kill them, so they
just wait for us to appear at the door or at the window. The officer
leading the operation always likes to be in the front. He is always

mentally prepared to take that first shot. His chaps, on the other hand, want their leader to be safe. But a true officer always leads from the front—and this is what happened to Tushar as well.'

The first squad sanitized the first floor and signalled the others to move to the second floor. Lance Naik Om Prakash of 9 Para (SF) was the leading scout of the second squad, which moved to the second floor during the building intervention. He was a veteran pointsman and, in the past, had been awarded the Asadharan Suraksha Seva Praman Patra in Special Group by the Prime Minister of India for eliminating four terrorists. He was also an excellent combat free faller. With many other accomplishments in Lance Naik Om Prakash's illustrious service life, he was always trusted with fulfilling the challenging role of the scout in any mission.

It was during the sanitization of the second floor that Lance Naik Om Prakash's squad drew heavy fire from the third floor. Realizing the mortal threat to his comrades, he immediately returned fire. This diverted the terrorists' attention but led to Lance Naik Om Prakash sustaining a grievous gunshot wound in the process. Heedless of his pain and blood loss, he defied the odds, advancing despite his disadvantageous position and displaying raw aggression and unparalleled courage. He eliminated one terrorist before being shot again. Lance Naik Om Prakash's incredible gallantry eased the pressure on his comrades and set the stage to eliminate the rest. Having ensured his squad's safety, he painfully extricated himself from the scene to take cover before being evacuated. Later, he succumbed to his wounds in the hospital.[18]

---

18 Lance Naik Om Prakash was later awarded the Shaurya Chakra for his indomitable spirit, his most conspicuous gallantry and selflessness while killing one terrorist, ensuring the safety of his comrades and extricating himself before making the supreme sacrifice.

This was a huge loss for 9 Para (SF), who were now desperate to eliminate the terrorists and avenge their comrade. The sacrifice of Lance Naik Om Prakash had not been in vain. The rest of the team knew the precise locations of the terrorists. Everyone else had taken cover during the firing. Until now, nobody knew exactly how many terrorists were present in the building—they would keep changing position and firing randomly, taking advantage of the stairwell and the concrete bunkers. This confused the men, and they decided to take all precautions to avoid any further casualties.

While approaching the third floor, Tushar's squad drew terrorist fire from one room. Undeterred, Tushar lobbed grenades and cleared the room that had now caught fire. Emerging into the corridor, the squad was pinned down by heavy fire from another room further ahead. Seeing his men threatened, Tushar advanced, firing and lobbing grenades simultaneously. A hail of terrorist fire caught Tushar in his legs, the force swinging him around midway. Bleeding profusely, he returned fire, wounding the terrorist. Shot again in the arm, Tushar displayed unparalleled resilience, aggression and raw courage. He charged and killed the terrorist instantly—before collapsing and succumbing to his wounds during evacuation.

Tushar's supreme sacrifice while safeguarding his men galvanized them into clearing the building and recovering the bodies of three terrorists. The operation went on for one more day because the forces still needed to sanitize each of the hundred rooms and ensure that there were no more terrorists hidden in the building or nearby. During the entire operation, many locals kept raising anti-national slogans and pelting stones at the same forces who had lost their comrades in the process of safeguarding the hostages inside the building. All the hostages were successfully released by 22 February.

Later, Gen Officer Commanding, Victor Force, Maj Gen Arvind Dutta, in a press statement, said that three terrorists had been killed and a large number of weapons and ammunition had been recovered.

The terrorists had apparently come equipped with enough provisions to stay for a longer duration. Dutta said, 'The terrorists had come to do damage to us, and there could have been much more collateral damage if the forces had not exercised caution.'

The operational brilliance of Capt Tushar Mahajan not only eliminated the terrorists within a day of 9 Para (SF) taking over operations but also prevented more casualties. A part of his citation clearly states,

> *Captain Tushar Mahajan was the architect of the Pampore building intervention operation wherein he formulated the plan, sited the fire bases and led the building clearance operation.*

~

22 February 2016
Udhampur
Jammu and Kashmir

Capt Tushar Mahajan's body reached Udhampur via helicopter the following day. A wreath-laying ceremony had been organized. Lt Gen D.S. Hooda, Army Commander Northern Command, was the first to lay a wreath and pay homage to this son of the soil. He was followed by other Army officers, the Deputy Commissioner of Udhampur and many police officials and politicians.

After that, Tushar's body was transported to his house in Udhampur. An enormous crowd accompanied the procession—people swore they had never before seen as huge a crowd than on the day the last rites of the true son of Jammu and Kashmir were performed. 'Captain Tushar Amar Rahe!', 'Bharat Mata Ki Jai!' and other patriotic slogans filled the air. Tushar's beloved tiranga swayed in the crowds, waved by people who had come to pay their respects to this son of the soil who had lived, breathed and ultimately died for his country. All business establishments remained closed that day, and his family and the unit received phone calls from across the nation. Tushar was ultimately cremated at Devika Ghat, Udhampur. The tragic pictures of the last rites when Tushar's mother cried her heart out over coffin of his young son went viral across the country, reminding people how freedom is not free.

Today, the youngsters of Udhampur join the armed forces to serve Mother India after listening to the story of Capt Tushar Mahajan, SC—much like how he did after hearing the tales of unimaginable bravery of the legends who came before him. Who knew that one day he too would be in the same league as them?

It was a tragedy for Tushar and Seema's families—they had been planning a beautiful wedding but now had to mourn the loss of their son. Tushar's parents, Dev Raj and Asha, were never able to recover from their unimaginable loss and eventually dedicated their lives to honouring the memory of their beloved son. Together with Shwetaketu, they established the Tushar Mahajan Memorial Trust to serve the people and nation, just as Tushar had done during his own lifetime. The trust helps to organize medical camps, educational scholarships, weddings of underprivileged girls as well as other social services for people in Jammu. It is also their attempt to heal the immense grief of losing Tushar much before his time. With the help of 9 Para (SF), the trust also erected a grand statue in Udhampur in honour of Capt Tushar Mahajan—a hero of the nation and the face of Udhampur. Everyone who comes to Udhampur visits his statue and pays homage to this brave soul.

~

This story is based on interviews conducted with Capt Tushar Mahajan's parents, Uncle Dev Raj, Aunty Asha and Tushar's childhood friend Shwetaketu Singh Jamwal. Aunty Asha broke down many times during her interview, but she insisted on telling me Tushar's complete story.

I am thankful also to Tushar's fiancée for sharing her most treasured and painful memories with me. Today, she leads a respectable and accomplished life.

Apart from his friends, family and relatives, I also interviewed around forty brothers-in-arms who cannot be named due to security reasons. These include Tushar's course mates from the academy, his fellow officers during the various missions and the soldiers who served under him. Each one of them came forward to honour his memory. These comrades of the maroon beret and the 9 Para (SF) have not left Dev Raj and Asha's side since Tushar's passing. They celebrate his birthday and veergati day every year with great bonhomie. Once a NINE, always a NINE—in life and in death. The purpose behind taking the arduous journey of writing this difficult story was to immortalize the valour of our bravehearts who went away too soon in the line of duty. It is our duty to keep their memories alive and tell our children freedom is not free. I hope countrymen would never forget there lived an enigma named Captain Tushar Mahajan, SC (posthumous).

5

Major Manish Singh, SC, and Aastha Panwar: The Love Story of a Commando

19 August 2020
Solan
Himachal Pradesh

TWENTY-SIX-YEAR-OLD AASTHA PANWAR was nervous, her heart was beating so loudly she feared people around would be able to hear it. She had left a letter for her mother back at her village, Deoriya, a distance of about an hour from Solan, the small town famous for its magical landscape in the lower Shivaliks. With its old church, ancient temples and captivating monasteries, this small Himachal Pradesh town had much to offer the many tourists flocking to it.

Aastha had only one thing on her mind. She was eloping. This pretty girl, with her beautiful eyes and smile, was about to run away with the love of her life, Major Manish Singh. He was waiting for her in Ghaziabad. They were scheduled to have their marriage registered in court the following day.

Manish had planned Aastha's escape. They had been in a relationship for a year by then. They had always wanted to get married, but her parents had refused. Manish had tried to make Aastha listen to her parents, but she loved him so much that she refused to do so. For her, it had always been love at first sight.

Maj Manish Singh, SC, was from 9 Para (SF) and also a Paralympian sharpshooter. He held an important position in his parent unit, where he managed the logistical support the unit required for its day-to-day functioning. His job required his complete attention. He would roam the entire campus in his wheelchair all day, getting his unit's work done. He is a living example of why SF personnel are so special.

It would have been easier for Maj Manish, SC, to give up and lead a life of anonymity—but he had chosen to serve his unit and nation until his last breath. He bled NINE in every sense of the word. As for Aastha, she had always been different. Her friends liked to party or go on picnics, but she liked the idea of devoting herself to the service of mankind. She was an ardent pupil of Sadhguru and had spent months serving in the Isha Foundation, a non-profit spiritual organization near Coimbatore. While her friends from engineering colleges cribbed about their jobs, boyfriends or the pressures of life, Aastha would be donating her hair to cancer patients and cooking and cleaning for people who visited the Isha Foundation in search of hope and peace. She did try to get a regular job and live like other girls her age, but she was unable to do so. Serving people would always be her main purpose in life. Her parents were equally spiritual, simple folk who lived in Solan, who had brought her up visiting ashrams and gurus. This lifestyle profoundly impacted Aastha, who grew up with a deep interest in spirituality and travelling.

When I met Aastha Panwar in July 2021, the first thing she told me was how she wanted to see the Northern Lights and travel all over the globe. There was a glow on her face—and I also noticed Maj Manish looking admiringly at his partner.

Maj Manish's story is the kind that can make anyone believe in miracles, in love and in fairy tales. Yet, his story also highlights the lesser-known side of the SF, the strength of the women behind the SF operatives. The power of the force behind the forces.

The story of this simple boy and extraordinary commando began in Ludhiana, Punjab.

~

29 February 1996
Ludhiana
Punjab

Layak Singh was upset with his younger son Manish's behaviour. The eight-year-old was throwing a tantrum—he was howling after a child's Army uniform that he had seen in the market and wanted for his second birthday. Manish, born on 29 February 1988, was a leap-year baby. Though they belonged to a humble background, his parents made sure to celebrate special days like this on a grand scale.

Unfortunately, even in 1996, that Army uniform cost Rs 400, almost three months' salary for Layak Singh. His doting wife, Sunita, was trying to persuade her husband to buy the uniform because Manish was such a good and obedient child otherwise. He was always the calmer, more tolerant, disciplined and mature of the two sons they had. But this Army uniform seemed to have captured Manish's imagination for a reason his parents couldn't fathom!

I remember Maj Manish laughing out loud at the memory. He told me, 'Ma'am, I have cried only twice in my life, and both times it was for the Army. The first time was when I was eight years old—for that olive-green uniform. And the second time I cried was when I was eighteen years old, and my parents would not allow me to join the Army.'

After finishing class twelve from Kundan Vidya Mandir, Ludhiana, in 2003, Manish's parents sent him to Kota for coaching for engineering exams. Manish was a bright student, and they hoped, like most average middle-class parents, that he would crack the IIT entrance examinations and lead a comfortable life. Manish worked hard in Kota, but something about the Army always pulled him towards it. He even cracked the All India Engineering Entrance Examination (AIEEE) in July 2006, and was admitted into one of the finest engineering colleges in India, the Birla Institute of Technology, Mesra, Ranchi (BIT Mesra).

His parents were ecstatic, but Manish, then seventeen years old, was troubled and confused. It was in the midst of this personal dilemma that he filled out the National Defence Academy (NDA) forms. There was something about the Army that called out to him like a siren. He knew his parents wanted him to be an engineer, but he just couldn't focus on his studies.

One Sunday, he sat for the NDA examinations and was subsequently called for the Services Selection Board. But he was in a real fix because he had been asked to report to SSB, Allahabad. If he went, he would need to inform his parents, and he knew they would be unhappy. Still, he called them.

'Hello, Ma! How are you?' said Manish.

'We are all fine, beta. How are you? I hope you are studying nicely. Papa has given his friends and relatives a treat. They wouldn't leave him alone; they kept saying now that your son is an engineer, why do you need to save money!' Sunita, replied with peals of laughter.

Manish's heart sank. He coughed a bit before replying, 'I am good, Ma. It is boring here. The college fest is going on, so all classes are suspended. I don't like these parties, Ma, so I was thinking of going to SSB, Allahabad, in two days.'

It took some time for Sunita to understand what her son was saying. After all, there had been many conversations how risky an option the Army was. Their hearts, as parents, were not strong enough to withstand the stress of having a son in uniform. They wanted him to opt for something safe, they wanted him to be an engineer—and he was already studying to be one.

Still, his mother didn't argue—especially after Manish cleverly explained that this would help him land a job in a high-end multinational company.

And so, Manish began his journey to SSB, Allahabad.

~

26 October 2006
SSB Allahabad
Uttar Pradesh

As soon as Manish stepped into the centre, he felt like he belonged there. The other defence aspirants, soldiers in uniform and the hustle-bustle of the place filled his heart with a joy he had never known before. The realization of his calling strengthened his determination to join the forces and wear the uniform.

He also knew it was a do-or-die situation. If he could not crack the SSB this one time, he would never be allowed to attempt it again. But he knew that the fauj was his true calling. He had to make it.

This realization filled him with josh. He had always been a studious boy. His knowledge of current affairs was already excellent, and he was blessed with an excellent physique.

The SSB went on for twelve days, and Manish kept clearing one hurdle after another—screening, physical or psychological and medical. Eventually, he secured the twenty-fifth rank in the NDA.

A huge smile played on Manish's lips as he told me the story years later. 'Ma'am, if you are destined for something, you will certainly get it. I remember I was asked to talk about "Whether women should be allowed in the infantry or not" in the group discussion. I spoke in favour of the proposition. In my interview, they also asked me about my hobby. I am a huge reader. Even today, I carry books everywhere with me. So, when they questioned me about my reading, I rattled off my favourites—from Charles Dickens to Shakespeare to Premchand. I have never once tried to be someone else. There was a josh in my heart, and I also had fun at the SSB centre. I made great friends there who were aspirants like me. I am still in touch with them. It was like the universe was in sync with me that time. It conspired to give me what I passionately craved.'

Manish quietly returned to his engineering college. He was there when he got to know about his selection on 29 November 2006. He was on cloud nine, but the immense pressure of convincing his parents to let him go suddenly dawned upon him. He reconsidered his options, but he knew he would never be happy being an engineer. So, the first thing he did was collect his documents from his college and leave the campus for home.

Manish's was due to join on 27 December, so he had some time at home. When they found him at their door, his parents thought he had come home for his semester break. When he told them the real reason for his coming home early, all hell broke loose. Layak Singh's anger knew no bounds; he felt betrayed.

'We sent you all the way to Kota. We spent so much money on you to become an engineer—and this is what we get in return? You have even withdrawn from the college without informing us?'

Manish was silent. His mother tried to intervene gently. 'Beta, we have seen the Kargil War. We have seen crying mothers, fathers and young widows. We don't have that kind of strength. Don't join the fauj. Be an engineer—stay with us,' she said.

'I tried to be what you both wanted me to be, but see what happened. I eventually did what my soul always craved. Don't stop

me, Ma. It will only complicate things further. I am happy, really,'
Manish replied calmly.

His parents knew then that they could not dissuade their son
from doing what he wanted to do. And since he had withdrawn his
admission from BIT Mesra too, they were not left with any other
options either. In a last-ditch attempt to cajole him, they asked
whether he would be able to complete the tough Army training in
the Academy.

Maj Manish laughs fondly at the memory. 'They had always been
a great pillar of support for me, but it was only when they attended
my passing-out parade that they finally truly accepted that they were
wrong and I was right. The grandeur of Army life mesmerized them,
and so did the charm of the uniform I was entitled to wear now.
From then till this day, nothing has deterred their faith in this great
institution, where people don't just do their jobs but also acquire an
alternate family which will never abandon you.'

~

27 December 2006
NDA, Khadakwasla, 117 Course
Pune
Maharashtra

With his first step inside the Academy, Cadet Manish was thrilled.
The manicured lawns, the awe-inspiring building, the hallways filled
with pictures of military heroes and the ambience suffused with
military culture and ethos introduced him to a wonderful world he
had never known. He was allotted to be a part of Echo Squadron, one
of the fifteen squadrons the NDA had at the time. These squadrons
can be compared with the houses in the Harry Potter books, where
each house has its own culture, values and legacies, and compete
fiercely with each other.

The Echo Squadron was famous for its sports records, but Cadet Manish was a studious boy who had never been good at sports. Amongst the elite Rimcolians and Sainik School students who had spent their entire life preparing for the fauj, Manish found himself lagging on the physical-fitness front. This resulted in frequent punishments that he or his overstudy received on his behalf.

One day, first-termer Manish broke down and cried his heart out. It was then that his overstudy, who was also allotted the position of his overall guardian, told him, 'You should not take anything to your heart. Rather, look for ways to improve yourself and bring yourself to the benchmark levels of this great institute, which has given this country so many heroes. We have given you punishments, but I don't remember any instance when we degraded you as a person. Do whatever breaks a man's body, but never break his dignity. If someone degrades you as a human being, that is wrong.'

Maj Manish told me that he follows this particular philosophy with his men even today. All the punishments and rigorous training schedules that all the cadets at the NDA followed toughened them with each passing day. The cadets would keep running around the campus and not find time to eat, sleep or even breathe properly. The harsh punishments, dispensed for no reason most of the time, would break Manish and the other cadets down, but they continued doggedly. Their overstudies would also explain to them how the purpose behind the NDA curriculum was to shatter the ego of each cadet and bring them all to one level. Any false ego was meant to be eradicated to create a true soldier.

The cadets' ustads would keep shouting, 'We only deconstruct you for the first three semesters so that you leave your previous identities. From the next semester onwards, we will start constructing you into an officer and a leader.'

The cadets were taught everything—to eat, drink, sleep, clean, make beds, live like a band of brothers, and everything else required to learn a holistic approach towards life. There were physical-training drills, exercises, parades, sports, and a lot more. Manish would struggle to keep up with this pace. He tried not to give up.

It was only in the fourth term, when Manish was introduced to the elite game of boxing, that his physical standards began to improve. Every cadet is expected to perform well in some games, but boxers hold a special place, and are even exempted from most punishments simply because the game requires the highest levels of mental and physical preparedness.

Cadet Manish became friendly with a Tajiki cadet during his boxing matches, who would keep encouraging him, 'Come on, Manish! What are you doing? If you cannot do it for your squadron, how will you do it for your country? Look at me. I have no affiliation to your country, but I still give my best for my squadron.'

That Tajiki was the best boxer in the squadron, and his words eventually changed Manish from a mediocre boxer to a performer. Maj Manish shared with me fondly how the foreign cadets who pass out from the NDA have so much loyalty towards the Academy and friendliness towards India forever in their hearts. Many of them also command the highest military positions in their home countries later and propagate the ethos of the NDA.

On 29 November 2009, after three years of rigorous training, Cadet Manish Singh took the Antim Pag (Last Step) at the NDA to join the IMA in January 2010, after a break of forty days. The IMA was a cakewalk for an ex-NDA person, and Gentleman Cadet Manish definitely excelled there.

In the IMA, he came across a Para SF officer, Maj Hooda, from the 9 Para (SF), who was their instructor. He inspired Manish, who was now filled with an insatiable urge to join 9 Para (SF). The Singhad company boy filled out the forms to volunteer for 9 Para (SF), and soon after his passing-out parade from the IMA on 11

December 2010, he left to undergo the dreaded probation period with his dream unit.

There was nothing Manish ever wanted more than to bleed maroon.

~

1 January 2011
9 Para (SF) Headquarters
Jammu and Kashmir

Lt Manish Singh stood at the gates of the elite SF battalion 9 Para, popularly known as the Ghost Operators of the Valley. The name was enough to send a chill down any terrorist's spine. It was now the sole aim of this young officer to become part of that legacy and don the coveted Balidan badge.

The three-month-long probation period was hellish. But it gave an opportunity for Lt Manish to witness men capable of being superhuman. The mental and physical agility of his ustads and every NINE he met was astounding. They seemed to be cut from a different cloth. The men were made to run speed marches, 100-kilometre marches; conduct demolitions; understand navigation; survive in the jungle alone and run up dangerous mountains with heavy loads on their backs.

Manish, for his part, performed supremely well. There was stress week when he was starved and beaten black and blue, and there were the escape and evasion exercises when he ran like a maniac from fake enemy camps. To join the 9 Para (SF), one needed not only to be physically and mentally fit but also have the highest levels of honesty and integrity. It was also important for the jawans to like their officer, and their opinions regarding the probationer officers mattered a lot.

In the end, out of the batch of fifty probationers, eleven were selected, out of which three were officers. Lt Manish was one of them. During the marooning ceremony, when the maroon beret was being placed on his head and he was gulping down whiskey, he knew he had achieved his biggest dream. This was his calling, and this was what he always wanted.

Immediately after the probation period, Manish was sent for real-time operations in the jungles along the LoC. Nothing was as easy as it looked. Just carrying weapons and walking with loads required a lot of skill and precision. One mistake could have jeopardized the whole mission. Random challenges appeared every second, from wild hounds to poisonous insects. The jungle was full of threats. There were times when Lt Manish treated his wounded comrades without any support in the deep jungles and snow-laden mountains. There were times when he mistakenly dropped his magazines, which were silently picked up by the men walking behind him. They were given back to him after a harsh ragda, or punishment, and a heavy briefing on how carelessness for even a second could lead to disasters. He also learnt tricks of the trade: ambushes, encounter, reconnaissance, surviving in the jungle and cold.

Manish loved his Tavor TAR-21, the Israeli bullpup assault rifle he carried with him as though it were a part of his body. It was also the first time he had been deployed to forward areas, which were snow-lined. The snow made a particular sound under Manish's feet as he walked under the canopy of coniferous trees. The sound reminded him that he was human, not just a soldier. On the other hand, the night-vision goggles that the team used during the pitch-black nights as they scanned the nearby areas for threats provided him with views from another world. It was a sort of spiritual awakening for Manish. There was beauty in solitude, but there were also life-threatening risks, such as deep crevasses. A single wrong step could be fatal. He had to be mentally and physically alert all the time. Sometimes, they slept on a branch of the tree or just under the starlit sky. Away from civilization amidst deep jungles, with its creatures and high mountains, he would feel as if life was simple. But a mere glance at his Tavor Tar and grenades kept neatly in his ruck sack would make him smile at the irony of the purpose of their presence in the valley. Sometimes, while waiting for a contact in a well laid ambush, he would remember his family and friends he had left behind. They would perhaps never know what all has gone on in the jungles to keep them alive.

In October 2011, Lieutenant Manish's Commanding Officer reminded him to complete his Para basic-jump course from Agra, which would equip him with new skills and entitle him to extra allowances.

He shared with me, 'Ma'am, parachute jumping for the first time is a spiritual and joyous feeling. When you leave the aircraft behind, you suddenly sense complete silence, which liberates you. You realize how everything in this world is trivial. You can sense your soul. You can compare it with the scene from the movie *Zindagi Na Milegi Dobara*, where Hrithik Roshan attains bliss after scuba diving. It's not joy—it's pure bliss. I feel everyone should keep their adventurous streak alive.'

When he returned from the course, Manish was a changed man. He also did some more courses required for a Para SF officer. He was then deployed to the general area of Kashmir, where the dynamics of the jungles changed entirely. It was not open terrain, where you could kill or be killed—now, there was politics involved. Manish saw how many locals tried to instigate the forces during their patrols to get an opportunity to make international headlines, but he applauded the patience and calmness that the forces showed in the Valley, even as they worked in tense civilian areas. They were also asked to maintain proper decorum while crossing by a religious place, in the name of respect. He learnt how to lay cordons and initiate search parties. He also acquired more knowledge on human intelligence and how a strong intelligence network has powers to change existing dynamics. He also realized that NINE has deep assets which have been inculcated over the years, and there was not one arena in the Valley where the unit did not have its own sleeper cells. The unit also had its covert operatives who led normal lives amongst masses but in reality were the fiercest SF operatives with immense skills that Nine had. Their lives were indeed the most challenging of all the operatives, but no matter what the cost, nation must survive. That the security and sovereignty of the nation come above all was deeply engraved on the heart of all the operatives.

For two years, Manish's skills kept improving as he participated in various operations. But he regretted not having any kind of direct contact yet. He kept his eye open for encounters, where he could prove his mettle. Be careful what you wish for, as they say.

~

25 September 2012
Somewhere in Kupwara
Jammu and Kashmir

On 24 September 2012, the 9 Para (SF) headquarters received a message from a Rashtriya Rifles unit about an ongoing operation. They were requested to be on standby in case more aggressive forces were required. Contact had been established with a group of foreign terrorists from Afghanistan. An intense firefight was on, in which the Rashtriya Rifles had already sustained one casualty while also neutralizing a lone terrorist.

The firefight had lasted the entire night—still, the terrorists had managed to escape. As a result, there was a grave threat of the terrorists joining their local sources and taking shelter in their safe havens. There was an equally dangerous chance of them crossing back over the LoC. The threat was considerable, especially since these terrorists were also equipped with the latest weaponry and plans to destroy India.[19]

In the morning, the elite 9 Para (SF) team was called was called to be the part of the SADO (search and destroy operation) of the Rastriya Rifles unit involved in this operation. Lieutenant Manish Singh was part of this group. The members of the 9 Para (SF) team divided themselves into three parties. Coordinates were issued, and command and control were established. Common communication channels were established with the Rashtriya Rifles team for the joint operation.

The terrorists were operating out of terraced maze fields where the line of sight was not clear. They were all well hidden and looking for an opportunity to fire, which happened later.

One party led an outer cordon to prevent any terrorists from escaping into the villages, and the two parties started acting as the main assault party and the static party respectively. The static party was supposed to give cover fire. They first decided to do a room-to-room intervention in the houses of the nearest village at the start of the jungle and terraced fields. It was a very stressful situation. As the soldiers went door to door, they had no idea which house was hostile and which door would open with a burst of bullets at them.

Lieutenant Manish Singh was leading his troops as a troop commander. He checked the ammunition in the pouch, his night-vision goggles tucked into his rucksack, which also contained a water bottle, rations, explosives, medicines and other items required during

---

19 The material recovered after the elimination of the terrorists revealed their plans. They had obviously been on a vicious mission to create the kind of large-scale national panic that would have created international headlines.

such search-and-destroy operations. It was a heavy load, but they were used to operating like this. His troop was tasked to track escaping terrorists through terraced maze fields.

The farmland was cut into different levels because of step or terrace farming, which made the operation difficult. While searching the area, the troops were visible to the terrorists, who in turn were hidden under the crops. The soldiers, fully exposed, could not see the terrorists, but the terrorists could see them.

Lieutenant Manish Singh asked his troops to crawl through the huge field, but they gave up after several hours. It was a meticulously planned operation, with different teams operating across their allocated locations. But after several hours, there was a kind of restlessness in the air. Manish had been crawling through the fields for a while. Dogs were barking in the distance, clearly indicating that danger lay ahead.

After hours of extensive searching and crawling, Manish eventually got to his feet. This increased the speed of the operation. The team soon found the footprints left by the terrorists by following the search dogs. The footprints vanished at the mouth of a nallah, but the search dogs wouldn't stop barking. Just as Manish was pondering a further course of action, several bursts of fire startled him. In a display of extreme courage and utter disregard for his own safety, despite his injuries, he crawled forward even after being shot and kept the terrorists pinned down. As one of the terrorists charged at him, Manish shot him dead at near point blank range before he slumped to the ground. One of the bullets had bored straight into his spine.

As the noise of the bullets died into silence, the Team Commander shouted, 'What happened? Is anyone hit?'

'Yes,' Manish said calmly, 'I have been hit.'

He stared up at the crystal-clear blue sky. He could feel the soil on his face, and though he tried to lift his hand to brush it off, he couldn't seem to move at all. He was thankful for one thing—his vision had not yet blurred.

Meanwhile, Manish's teammates were in a great dilemma. They had no idea of the number of terrorists in the area. Rushing the mission could jeopardize them all. So, they formed parties to evacuate their injured comrade and provide cover, along with searching and destroying. It took them a while to get to him, fallen as he was at the bottom of the nallah.

When they reached him, they scrambled together a stretcher using material from the jungle and some corrugated galvanized iron sheets. Manish was shifted to an ambulance before he was flown to 92 Base Hospital in Srinagar, famous for saving lives in the nick of time. Manish had, by this time, suffered massive blood loss and had passed out from the pain. But the doctors managed to remove the bullet and save his life. Since he was still in critical condition, he was airlifted to the Base Hospital in New Delhi, where another ten-hour-long operation ensued.

Manish had by now slipped into a coma. Doctors were unsure whether he would even survive. But the tough young commando put up a great fight and, within a month, was shifted out of the ICU. As Manish began to recover and come out of the coma, terrible nightmares—of blood, violence and death—began to torment him. It was traumatic to witness but it was, in reality, the sign of his brain rebooting after a long illness.

News of Manish's heroic encounter and his survival began to spread like wildfire, and people from around the country flooded the hospital to meet him.

Maj Manish Singh told me, 'It was as if the entire Indian Armed Forces knew about me and rushed to the hospital to meet me. There were veterans, newly wedded wives with their husbands, senior officials, generals and people from different walks of life coming to meet me. My parents never had a second to think about anything. Initially, when they saw me, my mother was shattered, but as soon as she realized that not just my unit but the entire fraternity was also standing behind me, she felt calmer.'

In December, Manish was shifted to Military Hospital, Kirkee, Pune, from Delhi. His buddy never left his side, and Maj Manish credits much of his recovery to him. After all, it was his buddy who had heard about the experimental stem-cell treatment being conducted in Mumbai and had, with great effort, taken Manish there and convinced the doctors to carry out the treatment pro bono.

He had told the doctors earnestly, 'My saab has fought for his country. He has suffered this bullet injury protecting you all. He did his duty honestly, and now it is your turn to save him. He does not have a lot of money.'

The doctors were also impressed with Manish's physical and mental resilience. The young commando told them, 'Do whatever you want with me, I have no problem. I only want to walk again and rejoin my unit.'

And so, the doctors at NeuroGen Brain and Spine Institute, Nerul, Mumbai, gave their best to Manish. He was kept under intense medical care. After stem therapy and working on bed-mobility techniques for several months, the brave officer was finally mobile, with the help of a wheelchair.

9 Para (SF), the unit famous for never leaving a brother behind, stood with him like a true pillar of support and provided him with everything he needed, sometimes going out of their way to help him. There is a reason they say, 'Once a NINE, always a NINE!'

On 26 January 2013, Manish was awarded the Shaurya Chakra for his exemplary courage and leadership. Later, it was revealed that while he had successfully eliminated one terrorist, the rest had escaped. They had been part of a new tanjeem that was trying to break through, but Lieutenant Manish's incredible bravery had broken the back of this new terror group. His citation reads,

> *For displaying courage of exemplary order and placing the safety of the men he commanded above his own, disregarding his own welfare and safety, showing nerves of steel and outstanding leadership in the highest traditions of the Indian Army, Lieutenant Manish Singh is awarded the Shaurya Chakra.*

Meanwhile, even though he was disabled, he thought greatly about rejoining his unit. On their part, the unit, too, would think about ways in which his training could be utilized in the best possible manner. It takes more than training to create one Para (SF) who bleeds NINE. Loyalty is everything—it is the core ethos of the unit.

There also came a turning point in Manish's life when he met India's star para-athlete, Deepa Malik. Seeing his mental strength, she asked him to consider sports. Manish discussed this with his Commanding Officer, and they shifted him to the Army Marksmanship Unit at the Infantry School in Mhow as an additional officer, where he would undergo treatment and practise shooting. At that time, Manish came into the media limelight for his exceptional shooting skills; he even wanted to participate in the Paralympics. But despite everything, he still missed his unit and his brothers.

I met Maj Manish in 2021, when he was still serving in the same capacity in his wheelchair. You and I cannot understand what being a NINE means. But ask any NINE, and he, like Manish, will tell you what it means to them, and how you remain loyal to the unit until your last breath. Manish's brother officers and the Commanding Officer understood this and called him back to the unit, where he was put on active duty. His job was to look after the administrative issues of the unit, which he did as he used to do. His physical condition could have meant staying home on a pension, but that was always the easier option. Maj Manish could never sit idle. A bullet to his spine, the disability in his body—none of those things defined him. The spirit and ethos of an Indian Army soldier made him who he is today.

~

July 2015
Wardha
Maharashtra

Twenty-six-year-old Aastha Panwar was reading *Pune Times,* when a news item on the third page caught her attention. It was about a Special Forces officer called Capt Manish Singh, SC. He was disabled and still his bravery knew no bounds. There are a few things we cannot ever fully understand—love, pain, admiration, entrancement, fear, joy, nostalgia, relief, romance and sadness. They are never in our control. Instead, they control us and decide our fate.

When I asked Mrs Aastha Panwar about it, she could not pinpoint what exactly had caused her to fall in love with Maj Manish Singh after reading that article. Yet, that is what had happened to her. People may have laughed at her at that time, but as you read her story ahead, you will see that true love is boundless; it doesn't follow rules.

Aastha reminded me of a story I had covered in my third book, *The Force Behind the Forces: Stories of Brave Indian Army Wives* (2021), where the same thing had happened to Sowmya Nagappa. She had fallen in love with Kargil War hero Capt Naveen Nagappa after reading a newspaper article about him and later married him.

Aastha, at that point, knew that she was destined to be with Manish. That was what pushed her to create a Facebook account and ping the Captain on his Facebook profile. He never replied. Little did Aastha know that he was posted at the Army Marksmanship Unit in Mhow and was engrossed in dealing with his life and shooting practice. Through Manish's Facebook posts, she realized that, like her, Manish was also a follower of Sadhguru. It was like yet another sign from the universe. But Manish would not respond to her. He thought no girl would ever be able to see beyond his disability.

When I met Aastha at their home in July 2021, the couple's first wedding anniversary was just a month away. She told me, 'I must thank Mark Zuckerberg for Facebook. It enabled me to connect with the love of my life. I knew I loved him. It was love at first sight. I would send him messages on Facebook. I wanted to say I love you at the first instance but since I knew he would not believe it, I changed it to I adore you.' Her sweet laughter filled the little fauji accommodation.

~

26 July 2018
Isha Foundation
Coimbatore
Tamil Nadu

Aastha Panwar was at Isha Foundation for several months. With her parents in Kuwait and her brother busy with his job, she thought it was the perfect time to serve humanity and drown in the ocean of spirituality.

On 26 July 2018, she opened up to her close friend Shailza about Capt Manish, telling her that though they had chatted through Facebook, Manish had never really shown an interest in her. Shailza encouraged her to start afresh and change her profile picture—which was then a pizza—to her actual face. The trick worked and piqued Capt Manish's interest during their next chat, where he found Aastha childlike and adorable. Mustering up a lot of courage, she asked him if she could send him a picture once she had had a haircut. Capt Manish said yes and soon found himself staring at the image of a bald Aastha. She had recently donated her hair for a ceremony at Sadhguru's ashram. She was utterly unaware of her looks; it had never mattered to her.

For Capt Manish Singh, it was a profound moment where he realized that here was a girl who was not superficial in any way; she was not fake and she did not pretend to be someone she wasn't. Aastha was a gem of a person who was least bothered about her external appearance (or anyone else's) and the material world. After all, she wasn't even bothered about her own baldness.

This is how the ice finally broke between the two. In August 2018, when Aastha completed her session at the ashram, she visited Capt Manish in Udhampur, at his insistence. He said, 'I know nothing matters to you except love, but seeing me in pictures and in real life are two different things. I insist you meet me before we decide our future.'

Aastha flew to Jammu and Kashmir to meet Manish. It did not matter to her in the least that he was disabled; she didn't even see it. Time passed like the wind. Aastha shared with me, 'Swapnil, for a long time, it did not register to me that Manish was in a wheelchair. For me, everything about him is whole. When he asked me about it, I told him that when you are in love with someone, you don't love them partially—you love all of them. I told him it comes naturally, and if I say "I love you", I mean it. Maybe we have known each other in our past lives. I have never felt that kind of connection with anyone else.'

Later, Aastha informed her parents about Capt Manish, and they panicked. They left Kuwait, returned to India and tried to dissuade her many times. But nothing deterred Aastha. She had always been like that. Once she decided on something, she would stand by it. In anger, her parents seized her phone. She was threatened. Her mother could not digest the fact that her daughter wanted to marry a permanently disabled person. But, as they say, love has the power to move mountains.

In November 2019, Aastha met Capt Manish's parents, who were also troubled. His mother even told Aastha, 'My son has already suffered a lot. I don't know if he would be able to take another heartbreak. So please say yes only if you dare to walk beside him all your life. It is completely your choice.'

Aastha just nodded yes. She knew what she wanted.

Meanwhile, the couple kept trying to convince Aastha's parents to accept their relationship, but nothing worked. So, one day, in the midst of the Covid outbreak, they decided to elope. The commando prepared a master plan—he named it 'Operation Love'. This time, the plan was to rescue himself.

Aastha was picked up by a 9 Para (SF) veteran from Solan and was joined by Maj Manish's mother at Pinjore, in the Panchkula district of Haryana. The two ladies reached Ghaziabad on 19 August 2020. It was Hartalika Teej[20] that day. They had purposely fixed the wedding date for 20 August because of this great festival, which was similar to Karwa Chauth for married women. Manish's mother observed that the story of Lord Shiva and Parvatiji matched her son's own love story.

And so, Manish and Aastha registered their marriage in court on 20 August 2020. This was followed by a traditional wedding ceremony on 21 August at Manish's parents' apartment. It was everything they had ever hoped for—though Aastha kept worrying about her parents. On 22 August, Maj Manish called her parents and informed them about their wedding. His parents also spoke to their in-laws and invited them home to Ghaziabad.

---

20 Hartalika is made of two words: 'harát' means 'abduction' and 'alika' means 'female friend'. The legend goes that Goddess Parvati performed severe penance at the bank of River Ganga in order to have Lord Shiva as her husband. It is mainly celebrated in Bihar, UP, and Jharkhand India, and also in Nepal, for the husband and for marital bliss.

Aastha's parents reached Ghaziabad the following day. When they met everyone in Capt Manish's family and saw how happy their daughter was, they relaxed and accepted the union wholeheartedly.

Maj Manish recalls fondly, 'Now, Aastha says I have stolen her parents from her. Every time she fights with me, her mother always sides with me. Since the day of our wedding till today, they have only spoiled me with their love and affection.'

When I asked the couple about their plans, Aastha said she was going for a trek in the Himalayas that December. She does not like leaving Manish by himself, but her love for travelling has remained alive too. She has big plans to travel across the globe. Scientific and technological advancements give the young couple a valid hope for having their own children, but not just yet. For Maj Manish, everything comes after his commitment to NINE.

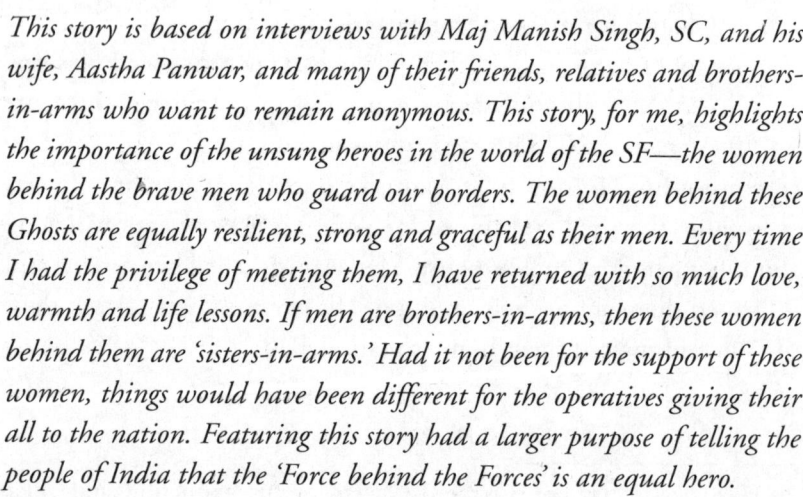

This story is based on interviews with Maj Manish Singh, SC, and his wife, Aastha Panwar, and many of their friends, relatives and brothers-in-arms who want to remain anonymous. This story, for me, highlights the importance of the unsung heroes in the world of the SF—the women behind the brave men who guard our borders. The women behind these Ghosts are equally resilient, strong and graceful as their men. Every time I had the privilege of meeting them, I have returned with so much love, warmth and life lessons. If men are brothers-in-arms, then these women behind them are 'sisters-in-arms.' Had it not been for the support of these women, things would have been different for the operatives giving their all to the nation. Featuring this story had a larger purpose of telling the people of India that the 'Force behind the Forces' is an equal hero.

4 Para Special Forces: The Fourth of the North

THE 4 PARA (Special Forces) call themselves the 'Fourth of The North'. 'Daggers', 'Saviours of the North' are other popular names used by the masses. They also call themselves the 'Gentlemen of the Special Forces' as they are known for their fierce and professional operations and minimal casualty rates. They were thrust into the limelight after the surgical strikes of 2016. The world was in awe, and every Indian talked about the lethality, speed and precision of the cross-border operation. The destruction in the enemy camps and the damage inflicted looked unreal. In fact, it was not the first time that 4 Para (SF) has been involved in a daredevil mission across the borders. I was told that the unit plays a role in every operation that is deemed to be dangerous or impossible across the border at the LoC.

They do their job quietly. The Fourth Battalion, the Parachute Regiment, was raised on 1 August 1961, in Agra, and the orders to be converted into a Para (SF) unit were issued on 11 September 2003 and completed on 30 June 2004. The unit has the unique distinction of being the first-ever battalion of the Indian Army to be originally raised as an airborne battalion. The unit has collectively been awarded one Maha Vir Chakra, three Kirti Chakras, two Vir Chakras, twelve

Shaurya Chakras, sixty-five Sena Medals, thirty Mentions-in-Despatches and sixty Chief of the Army Staff Commendations, along with being recipient of many other awards and honours, including battle and theatre honours in a relatively short period, as compared to other SF units.[1]

The battalion was earmarked for special missions along the northern borders of India, which is the most active operational area. From Operation Vijay for the liberation of Goa in 1961 to Operation Leghorn in 1962, from the Cho La defences in 1965 to the famous Operation Cactus Lily in 1971, and from Operation Pawan in 1987–89 to Operation Meghdoot in 1994 and the punitive strikes on terror launch-pads in 2016, there are many accomplishments to 4 Para's name.

The battalion, famous for its low casualty rates—despite being involved in various operations throughout the year in a volatile zone—had an unfortunate operation in 2020, when they lost five of their comrades in Operation Rangdori Baihk.

However, when I started researching the subject, I realized that the operation was a burning example of camaraderie and raw heroism. It also showed us how sometimes losing is more important than winning, and victories are not defined by medals but by the men who fiercely fought, believing in each other to protect every inch of their motherland.

> *To every man upon this earth,*
> *Death cometh soon or late.*
> *And how can man die better*
> *Than facing fearful odds,*
> *For the ashes of his fathers,*
> *and the temples of his Gods?*
> —Thomas Babington Macaulay, 'Horatius',
> *Lays of Ancient Rome*

---

1 Based on data available till 2022.

6

Operation Rangdori Baihk: The Band of Brothers

6 April 2020
Operation Rangdori Baihk
Shamshabari ridgeline
Kupwara sector
Jammu and Kashmir

IT WAS HOSTILE weather. The temperatures had fallen to such lows that stalagmites of ice had formed on the mountains. Blizzards were raging, and the visibility was below zero. This was the land of the forsaken, with its steep mountains, narrow gorges and deep jungles of coniferous trees sprawled across the sheer white expanse of snow.

Yet, that morning, the unmanned aerial vehicles (UAVs) flying silently across the terrain highlighted nine entangled dead bodies. Crimson-red blood was splattered all over the pristine white snow, lending the landscape a grotesque appearance. A closer look confirmed close combat between the soldiers of the Indian Army and the terrorists across the border.

These Army personnel were no ordinary soldiers or commandos. They were the most elite SF operatives, belonging to the 4 Para (SF), referred to as 'the Fourth of North'.

This, then, was the result of no ordinary fight but of close-range, hand-to-hand combat, with bullets being fired at point-blank range, rare for encounters in Kashmir. For the comrades who reached the encounter site at first light, it was a heartbreaking sight to witness, a huge loss.

In the glorious history of 4 Para (SF), the unit had never before faced such a casualty, despite handling risky and stressful missions with immense challenges.

There were the dead bodies of the bravest of the braves, included Subedar Sanjiv Kumar, Havildar Devendra Singh, Paratrooper Amit Kumar, Paratrooper Bal Krishan and Paratrooper Chhatrapal Singh. Initial observations suggested broken bones and limbs in two of the soldiers, while the bodies of the other two were riddled with bullets. Yet another body was found entwined around the body of a terrorist.

But one question continued to ring in the minds of the men: *What had happened here?*

~

1 April 2020
Keran sector
Kupwara
Jammu and Kashmir

It all began on 1 April 2020, when unmanned aerial drones monitoring the LoC sent the command centre some disturbing images of several men infiltrating the Juma Gund area in the Keran sector. For the men of the Army, the onset of summer means harvest season is around the corner, the harsher winter paving the way for softer snow. It is during this period that the heaviest infiltration is reported, with trained terrorists crossing the border from their launch pads across the border to reach their designated locations in the main valley on the other side of the border. The challenge for the terrorists is to dodge the Indian Army deployed in the area before establishing contact with local OGWs and entering cities with their plans of destroying peace on Indian soil.

One of the SF officers jokingly said to me, 'But Ma'am, we welcome them wholeheartedly. How would a farmer feel about not cutting his crops? The harvesting season truly delights us when they come in a bunch, ripe and fresh, providing us with the opportunity to cut the crops! After all, we hone our skills the entire year just for the harvesting season.'

This mischievous quip was followed by a peal of laughter, but as the story progressed, the lightness dissipated in the face of the grief he still feels at the loss of his brothers-in-arms.

That day, at the divisional headquarters, the intelligence input for the possible infiltration consolidated the evidence of the existing surveillance and patrolling mechanisms. The anti-infiltration obstacle system (commonly referred to as the AIOS) was buried under at least 7–8 feet of snow due to constant snowfall over the past few days. The deployment of the troops was beefed up after the infiltration input to ensure the aggressive domination of the infiltration routes. UAVs were extensively used for the area's surveillance, and many helicopter sorties were conducted to aerially track the infiltration. These measures were tremendously beneficial in keeping the terrorist movement under control, pinpointing their likely location and effectively cordoning off the area.

It was at this point that an infantry unit deployed in the area found footprint trails. They followed them to locate the terrorists far up in the hills, hiding in between the trees. A firefight ensued at around 1 p.m.—it was brief, and contact with the terrorists was not so successful. Still, in their haste to save their own lives and escape, the terrorists threw away their rucksacks. When these were recovered by the troops, they revealed essential items for sustenance—GPS trackers, laminated flaxseed maps, satellite phones, medicines (like morphine, blood coagulants, various energy and painkiller injections), several packets of dry fruits, dry ration and rotis and extensive ammunition including Chinese grenades, American night sights, several magazines, etc.

One of the personnel involved in the operation told me, 'Ma'am, it was a desperate situation for the terrorists. They were trapped. Their location was cordoned off, their rations had been captured, and low ammunition made them distressed and dangerous. It was a do-or-die situation for them. We also had pressure on us to track and eliminate them soon, because chances were that they could have crossed the LoC back to their launch pads.'

The terrorists by then had slipped into a deep gorge in the Rangdori Baihk area, famous for its steep mountains, dangerous nallahs and haphazard patches of coniferous jungles and icy snow. It was a rugged terrain to operate in because of the natural barriers and obstacles which hindered them from a clear view of the terrorists. In comparison, the terrorists had excellent hiding spots in the vast terrain. So, the area was soon cordoned off for search-and-destroy operations (SADO), as soon as the information regarding the failed contact flowed into the divisional headquarters.

As part of the initial response, the following day, the nearby villages of Auwoora, Kumkadi, Zurhama, etc., were surrounded by the Indian Army and cordoned off to prevent an escape or retreat by the terrorists. On 2 April 2021, two different units of Rashtriya Rifles also joined the operation, and they were successful in establishing contact with the terrorists in the evening. It was an intense but brief contact from both sides, but the heavy snowfall, low visibility and rugged terrain provided excellent cover to the terrorists and proved a blow to the second contact as well to the Rashtriya Rifles. The terrorists were successful in dodging the troops and sliding deep inside the jungle. The situation at headquarters was tense, and the troops' desperation to eliminate the terrorists was palpable.

The various brigades deployed in the area started setting up stops at probable escape routes. Various battalions began patrolling the area, and UAVs constantly flew overhead, trying to spy on the terrorists. Two more contacts were established in between the search but each failed. Conferences for effective SADO plans at the brigade headquarters continued. It had now been four days, and the Brigade Commander was worried about the terrorists retreating to their launch pads or creating serious havoc during their last-ditch attempts to escape. It was at this moment that the SF were inducted into the operation to eliminate the terrorists. It was decided to call in the elite 4 Para (SF), who specialized in mountain warfare.

The two squads[1] from 4 Para (SF) flew to the brigade headquarters, where they were briefed on the latest feeds from the UAVs and provided with marked maps of the area with the probable presence of the terrorists. While the location on the maps was accurate, the demarcated area was vast and completely covered in a white blanket of snow, with patches of sharp coniferous jungles hiding steep gorges and giant boulders. The SF had been deployed a little late, but they were confident of accomplishing their mission under the command of Subedar Sanjiv Kumar.

Subedar Sanjiv Kumar knew the area like the back of his hand. In his twenty-four years of service, apart from a three-year tenure in the National Security Guard, he had mainly been posted in the

---

1 Squads is equivalent to a section in an infantry battalion, with a different strength than infantry subunits. There can be various squads under a troop (equivalent to platoon), which comes under a team.

Valley, from where he served the nation dutifully. The tall, lean and sharp Subedar Sanjiv Kumar, a native of Bilaspur district, Himachal Pradesh, considered Kashmir his second home. His wife, Sujata, would smile when he would wax eloquent on the beauty of Kashmir and Kashmiris every time he came back home on leave.

He would say, 'Sujata, not all Kashmiris are antinational or stone-pelters. Many are patriotic, and they want peace and prosperity in Kashmir. This hurts the sentiments of our enemy nation and its intention to disrupt growth and development in Kashmir. So, they keep on sending their boys to our land—but your husband and his brothers sit right there at the right spot.'

His remarks would always be followed by a burst of laughter. The fair and tall Sujata, who I met in November 2022, shared how she can still hear his laughter even after two years. She said, 'My husband was a proud man—he was first a soldier, then a brother to his brothers-in-arms and, finally, a family man. The family was last in his list of priorities.'

Sanjiv was the most dependable junior commissioned officer in his unit. That is what prompted the Team Commander to relax after issuing orders. The battle-hardened Subedar saab was all set to lead the squad. He briefed his seniors on his excellent plan to search and eliminate the terrorists. After getting quick approvals, he boarded the helicopter, ready to travel to the operational area. It was a simple plan, and he had had years of experience executing similar missions. After all, he had been part of numerous complex operations before and had a record of never losing contacts. In his mind, everything was in control.

~

4 April 2020
Operation Rangdori Baihk
Keran sector
Kupwara
Jammu and Kashmir

The twin-engine, multirole, multi-mission, new-generation Advanced Light Helicopter (referred to as ALH-DHRUV) took random rounds around the marked landing zone. This was done to ensure that the pilots did not drop the troops into direct firing lines, considering the threat of hidden terrorists in the area. There might also have been an ambush laid in the area. As a result, the pilots performed standard confirmatory recces while the cockpit heated up with last-minute preparations. The SF operatives, on their part, were gearing up with weapons, equipment and snow-appropriate clothing.

The men suited up in specialized white windcheaters, which provided warmth against the freezing temperatures as well as being excellent camouflage. Specially designed ballistic anti-glare snow goggles were also packed. The scouts checked their thermal imagers one last time. It was their duty to find a clue and inform the squads. They also packed in their radio sets, medicines, rations and water in various pouches.

The array of personal weapons that each of the SF operatives loaded onto themselves was massive: they had Pika guns, under-barrel grenade launchers, Tavor TAR-21 assault rifles, M4s and grenades. The pilots signalled 'all okay' for the 'heli-dropping' of the SF's detachment. The leading Squad Commander, Sanjiv saab, checked his map and GPS one last time before signalling scout number one, Paratrooper Chhatrapal, to jump off the heptr, which was now hovering steadily above the marked landing strips.

The huge rotors were spinning, and the snowstorm swirling outside was causing the helicopter to wobble slightly. Paratrooper Chhatrapal jumped out of the helicopter and sank straight into the soft snow. Fortunately, his head was still clear of the snow, and he immediately undertook one of the avalanche drills that he had practised so often before to break loose. Sanjiv saab, whose prime responsibility was to ensure the safety and security of his men, was alarmed.

Sanjiv saab informed the pilots about the situation and requested them to drop the remaining soldiers into another safe zone nearby. The pilots complied, and soon the two squads were heli-dropped safely. They waited for Paratrooper Chhatrapal to join them, and once he walked up to them, Sanjiv saab briefed the squads one last time and established a standard formation, trailing the most probable route in search of terrorist footprints.

Twelve men were on the ground, which was covered in 8–10 feet of snow. Walking through it was next to impossible. They waded through waist-deep snow with heavy battle loads on their backs and weapons in their hands. They proceeded cautiously. Though they were trained operatives, they knew that the terrorists could be waiting anywhere to ambush them. It was difficult not to notice the enormous helicopter in the serene and still valley. After wading through heavy snow for five hours, Paratrooper Chhatrapal, the lead scout, suddenly gave a field signal to halt the squad. All twelve men immediately fell to the ground, submerging themselves in the snow. They became invisible.

Everyone could see a cluster of coniferous trees ahead, which could have been the likely location of the terrorists. The highly skilled Chhatrapal had spotted two silhouettes amongst those clusters of trees.

Chhatrapal was a boy from Jhunjhunu in Rajasthan, where there is a soldier in almost every house. His mother, Shashikala, had wanted him to become a driver, but he always told her that the Army was his passion. His passion for fighting and leading an adventurous life eventually pushed him to join 4 Para (SF) in 2018 from the Indian Army Service Corps. When I met his mother, I found her to be a strong woman. No wonder Paratrooper Chhatrapal had inherited a fearless attitude. Shashikala, who lost her son too soon, sometimes feel as if her son would enter her house and put his head on her lap as he used to. Chhatrapal one of the top probationers of his batch, where the attrition rate was 99 per cent. Mediocrity was not something he liked; he believed in excelling. He was curious and chased his ustads to explain concepts to him if he did not understand anything.

Today, his friends laugh and say that if the handsome light-eyed Chhatrapal had not joined the Para SF, he could have made it big in Bollywood for sure. He liked John Abraham a lot, and he was a gym freak with a passion for bodybuilding. He was meant to do great things in life, but nobody knew that he would record his name on the unit's honour board this soon.

Until now, the terrorists were unaware of the Army's presence. Now, it was time for scout number two, Paratrooper Bal Krishan, to perform. He crawled through the snow, confirming the presence of two terrorists. He signalled silently to Squad Commander Sanjiv saab, who was present at number three in the formation

It was a tough call. The soldiers had the advantage of a surprise attack, but that would last only for a few seconds. After that, it would be a free-for-all for both parties. To maintain the surprise element, they would have to fire from a certain distance. The slightest sound in the silence would be enough to give their position away and alarm the terrorists.

Sanjiv saab was a father figure for his boys. He did not want to take the slightest chance regarding the safety of his men. The experienced JCO divided the twelve men into two parties, to close in from both sides. The first squad would act as the main assault team to attack from the front, while the second squad was tasked to climb onto the nearby heights, dominate the position and open fire as the support team when required. The tactical experience of operating at such terrains throughout his military career was coming in handy for the hardy Sanjiv saab. In a firefight, either squad would always be free from range, and would be able to freely manoeuvre, even if the other squad was pinned down.

Sanjiv saab signalled the start of the operation. While the main assault team started crawling and deploying their weapons tactically, the other squad started leapfrogging.[2] The operation had begun four days ago, and the 4 Para (SF) had taken a mere five hours to track the terrorists. The men desperately wanted to eliminate them and prevent further tracking and trailing.

The moment Sanjiv saab signalled his men to fire, heavy firing erupted at the terrorists sitting cosily at the tree roots, but not one bullet could touch them. The lost moments of opportunity were enough to alert the terrorists. Their position was desperate, and they knew it. They were like deer caught in the headlights of an oncoming car. Darkness had fallen, and under the cover of night, the terrorists escaped even as the men kept firing.

After an hour or so, an eerie silence fell in the area. Sanjiv saab asked his men to stop firing and asked some chaps to check on the terrorists. The men soon confirmed that the terrorists had escaped. This was disheartening for the squad, but they managed to lay their hands on vital injections, a pair of gloves and some medicines while searching the area. The squad located the trail of the terrorists too. It seemed that they had jumped further downwards towards an existing nallah.

---

2   A tactical movement when troops are not moving at an equal pace to prevent firing on everyone at once, while gaining complete freedom over heavy suppressing fire

The troops knew it was only a matter of time before they eliminated the terrorists. The starved and depressed terrorists were surrounded on all sides, and the luck which had favoured them so far had to run out sooner or later. It was just a cat-and-mouse game at this point.

Subedar Sanjiv Kumar gathered his men again and briefed them about the current situation. The terrorists knew that they had been spotted. They also knew that there were now troops in the area—and so they might as well be lying around with ambushes in wait.

Sanjiv saab took out his radio set and established contact with headquarters, saying, 'Delta to HQ, Delta to HQ. Contact established, two terrorists confirmed, more could be around. They have moved further downwards, alert infantry posts to put a stop. Over and out!'

The temperature had fallen by several degrees and the snowfall had intensified. It had now been eight hours since the men from 4 Para (SF) had been on the chase. By then, they were wet and shivering. They could have easily rested that night and begun their search again the following day. But the duty-bound SF operatives wanted to leave no chance for the terrorists to escape. They decided to give it one last shot.

For the next four hours, the SF team followed the footprints trail until the nallah, but they lost track after that. The squad knew that since they had sanitized the area behind them and that stop posts were placed in the area ahead, the terrorists should be hidden around somewhere in this area.

In front of them, an enormous canal with a 60-degree gradient flowed ferociously. It was huge, with gigantic boulders nearby. Several trees had also fallen in the area. There was no soil, just a rocky mountainous canal with two steep spurs[3] protruding outwards. The

---

3   An area of high ground that sticks out from a mountain or hill

entire area was rigged with razor-sharp ridgelines. To clear the nallah in almost zero visibility was a risky nightmare.

However, battle-hardened Subedar Sanjiv Kumar was relentless. It was not the first time that he had faced an impossible situation. Again, he laid a tactically brilliant plan and divided the party into two squads occupying two spurs at a height that would give them dominance over the area. He also briefed his men to wait attentively at those spurs throughout the night and not move until he signalled that it was safe to do so via his radio set. Losing that position could have been dangerous in the open terrain. With only twelve men at his disposal, Subedar Sanjiv Kumar beautifully created an arc of defence with them, which would have ensured their survival through the night, with complete dominance of the area till daylight—which was when they could have entered the nallah and continued with their SADO. According to any war manual, this was a brilliant plan. There was not even the slightest chance of any mishap, and the troops all felt confident under Subedar Sanjiv Kumar's able command.

Sadly, what humans tend to forget is that nothing is really in their hands—sometimes even the most perfect plans can end in disaster. And so it was that the operatives set out to occupy the spurs, forming two squads in perfect formation, in order to lay out a brilliant cordon.

The second squad—which comprised Sanjiv saab; Havildar Devendra Singh, who was also their medic; the lead scouts, Paratrooper Chhatrapal and Bal Krishan; the PIKA boy, Paratrooper Amit Kumar; and one more paratrooper who was the only one to survive—did not realize that the spur was not solid but a hollow cornice, an overhanging patch of hardened snow with no underlying support. It was right then that the cornice broke under the weight of the leading scouts Chhatrapal and Bal Krishan, who were walking ahead of the rest of the squad. They fell right into the abyss, tumbling down a total of about 150 feet.

One of the 4 Para (SF) boys shared with me, 'Imagine falling from that height. It was pitch-dark, the boys had been involved in strenuous activities for the past eight hours in the wet and cold, and then they fell from that height at such a steep gradient. It was heartbreaking. No mishaps had occurred over the first eight hours of the operation. And right when they were in the position of laying a comfortable ambush for the night, they fell off the ledge—and that too right in front of the hidden terrorists. Their fall would have broken bones in their bodies—so they were not even in the position to move and defend themselves. Also, on the fateful night, the terrorists were also hiding at the exact spot. They could have been anywhere else, and we could have easily retrieved our boys. Calling the incident unfortunate is an understatement.'

The X-ray reports of Paratroopers Chhatrapal and Bal Krishan showed several deep fractures, and it was the sheer resilience of their muscular bodies that they even survived the fall. The family of Paratrooper Bal Krishan still finds it hard to believe that he is no more, especially his brother, who is a soldier himself. He was posted in the Rashtriya Rifles when he received the dreaded call from 4 Para (SF). His first reaction was disbelief. He said, 'You are lying, saab. Bal Krishan is a commando. Bullets can't touch a commando. You are lying. Please check the facts. This is a terrible joke.' When I met Paratrooper Bal Krishan's mother Indira Devi, I realized how similar he looked to his mother. During the course of our interaction she kept on sobbing; her grief knows no bounds. Bal Krishan was the son she loved the most and she received equal love and respect from him. The loss was huge; a mother lost the apple of her eye so that Mother India remains.

The quiet, reserved and studious Bal Krishan aspired to become a medico one day. During his probation period in 4 Para (SF), he had chosen medicine as his specialized skill. He often made diagrams of the human body and pasted them on the walls of his room so that they would be the first thing he saw when he opened his eyes in the morning. Even though he was an SF guy, he had great compassion for people.

One of Bal Krishan's close friends shared with me, 'Madam, you know, once, during probation, he came across another probationer who was drinking water from a nallah. Probationers are often not given water, and to satiate that kind of thirst, people tend to drink water from anywhere, even though it is forbidden and a punishable offence. One of the ustads randomly saw someone drinking water from a distance and called both boys to ask who was drinking dirty water. Bal Krishan was such a saint that he took all the blame on himself even though the other chap did not open his mouth once. Bal Krishan's sense of integrity saved him, and he was eventually selected for the 4 Para (SF) unit, while the other guy was rejected.'

Paratrooper Bal Krishan was compassionate, and his endurance levels were also high. His fellow soldiers remember how during the 100-kilometre speed march, his legs were bruised entirely, almost broken, and yet he managed to finish the march on time without complaining about his pain even once.

Perhaps it was this endurance in the quiet and reserved man from the Kullu district of Himachal Pradesh that sustained him during his last hours when, despite being grievously injured, he was able to pick up his gun and fire at the terrorists.

After both the paratroopers fell off the ledge, they were mercilessly fired upon by the terrorists. It was at that moment that Subedar Sanjiv Kumar and Paratrooper Amit Kumar heard the firing, even as they were on their way to evacuate their comrades. Because it was pitch-dark, they could not see the terrorists and shouted in shock, '*Fire kahan se aaya* (Where did the firing come from)?'

Paratrooper Bal Krishan had shouted back, 'The terrorists are here! We are being fired upon.'

Imagine the plight of those two brave boys who were so battered and bruised and facing heavy fire—and still the brave lads of Mother India refused to bow down and fought till their last breath. With their broken bones, lying completely exposed, without the cover of a tree or boulder, they tried their best to aim and counterfire at the terrorists who were taking cover under a huge boulder. Clearly, the SF create real-life supermen.

It was now time for Subedar Sanjiv Kumar to retake command. He could have waited for reinforcements or fired undercover. He knew if he went out there in the open in the middle of a firefight, he was risking his life and making himself an easy target. But he jumped down nonetheless—for him, his boys were the most important.

Close on the heels of Subedar Sanjiv Kumar, Paratrooper Amit Kumar also jumped, following the lead of his commander. But because Amit was at some height, he placed his PIKA with some support on a rock out in the open, sustaining some bullet injuries while doing so.

Paratrooper Amit Kumar, a boy from Pauri Garhwal in Uttarakhand, was an excellent runner and had recently won an ambitious SF cross-country race for his unit, where every competitor was already the best of the best. He was originally from 16 Garhwal Rifles. He joined 4 Para (SF) in 2012, intending to serve the country in the best way possible. Amit's ustads remember his IQ being as high as an officer's—and his ever-smiling face. That smile would never fade, even in the toughest of situations. Amit was lean and athletic.

His mother, Bhagwati Devi, has lost the zeal to live. She broke down many a time, tears streaming down her eyes, which she kept on wiping with the help of her pink saree. I could not ask her anything. My throat was choked with emotion for a long time, even though I tried to smile before her.

Amit's stamina and fitness made him the perfect PIKA boy who could carry the heaviest of weapons and walk for kilometres during operations. Perhaps this was the reason why Amit Kumar did not just follow his commander but also shot suppressing fire, displaying his best firing skills—despite being shot himself. He was firing out in the open and would be an easy target. This did not deter him—he was undaunted and fearless.

Someone was shouting over the radio set, '*Saab aage mat jao, halt karo, we will tackle the situation* (Sir, don't move ahead! Halt! We will tackle the situation).'

But Sanjiv saab did not pay heed. Unmindful of his own safety, he pushed ahead into the nallah while Paratrooper Amit provided covering fire. Sanjiv saab dragged one of the scouts from the contact site to a distance and moved forward to retrieve the other scout, when he came under intense fire from the hidden terrorists. He realized the imminent danger to his squad members, crawled towards one of the terrorists and engaged him in hand-to-hand battle.

Meanwhile, Amit also sustained many bullet wounds, but he kept on firing. Havildar Devendra was behind the squad and could sense the danger, but the darkness and distance gave him no clarity. He was also the only one with a radio set, and he informed the other squad about the contact. But since the terrorists had captured the personal weapons of the fallen scouts and started firing from their Tavor TAR-21[4] instead of their own AK-47s, it confused everyone awaiting orders from Sanjiv saab. Everyone was utterly unaware of the situation he was in.

By then, Havildar Devendra had reached the height where Paratrooper Amit was still firing from, even though he was not in a condition to speak. Havildar Devendra kept on shouting, '*Saab, kya ho raha hai? Kya ho raha hai? Aadesh do* (Sir, what is happening? What is happening? Give me orders)!'

---

4   Tavor Tar-21 is popularly used by Indian SF operatives, while the AK-47 is mostly used by terrorists.

No reply ever came.

Havildar Devendra was also shot with several rounds during his attempt at evacuating Amit and bandaging him with supplies from the medical kit. In between all this, Devendra fired back when he could.

All five men were shot by several rounds as they tried to save each other. They didn't think once about their own individual safety. All their energy was directed towards helping their comrades.

The sixth member of their squad, Paratrooper X,[5] finally had a window of opportunity to enter the nallah, retrieve Sanjiv saab and shoot the other terrorist at point-blank range with his gun. One of the five terrorists was escaping, but Paratrooper X also threw a grenade at him. Some of the resulting shrapnel hit him too. Though the terrorist managed to escape, he was shot by the same infantry soldiers the following morning.

Sanjiv saab, Amit Kumar and Havildar Devendra had managed to shoot the rest of the terrorists. Whoever was left untouched was doubly shot by Paratrooper X, the only surviving member of the squad. He received the Shaurya Chakra later for the exceptional bravery and calmness he displayed even as the rest of his squad members had attained veergati, the supreme sacrifice in any brutal operation. It was not an easy situation for him. Soldiers are taught to work in buddy pairs and follow orders. Those circumstances could have overwhelmed any normal soldier but they didn't break him. There is a reason why we call them Special Forces.

The squad at the other spur kept waiting for orders from Sanjiv saab, which never came. There was no radio contact, and without any information, leaving their position could have jeopardized a well-laid-out tactical plan. They also feared mistakenly firing upon their brothers in the darkness. They kept hearing constant

---

5 The unit wants the sixth paratrooper of the squad to remain anonymous.

Tavor shots, which made them think that their brothers were probably successfully pinning down the terrorists, and that if any reinforcement was required from them, the radios would crackle with the request.

They told me sadly, *'Ma'am, humara kaleja strong hota hai, par itna nahi hota ki apne sathi ko marne ke liye chhor de* (Ma'am, we are courageous, but we aren't so courageous that we can leave our comrades alone to die).' They were referring to Sub Sanjiv Saab who, despite being aware of the dangers ahead, could not stop himself from trying to save his fallen comrades. The rest of the squad members decided to die fighting in their attempt to save their brothers.

There were two radio sets—one was with Sanjiv saab, while the other was with Havildar Devendra. Sanjiv saab could not communicate because he was trying to save his scouts. A hand-to-hand battle ensued. At the same time, Devendra was too close. He initially contacted the other squad when he decided to go down but later in the thick of the action, when he was fighting alongside Paratrooper Amit, who was trying to provide covering fire to the rest of the squad in the nallah Devendra could not. Imagine the scene: brave injured soldiers spending their last breaths trying to safeguard their brothers-in-arms. Saving the brothers with vengeance would be the only thing in their minds, I suppose.

We talk about Krishna and Sudama's friendship, but bigger examples of camaraderie or brotherhood exist in the Indian Army. Can we imagine the aggressiveness and passion these men displayed while fighting in order to save their comrades? What kind of men don't think about their families but sacrifice their lives for their comrades and nation? Tears were rolling down my cheeks by this time while listening to this extraordinary story—the officer's eyes were moist too. The desperation to save their comrades, while putting their own lives on the line, was a display of the highest level of military bearing, and utterly heartbreaking to witness.

The UAVs had been flying around all day on 5 April 2020, before returning to base around 7 p.m. After their flight the following morning, they returned with heart-wrenching images.

And what about the other squad? They were shell-shocked when they reached the contact location down and reported, 'We kept hearing firing, but since it was a Tavor (which had been captured by the terrorists), we thought it was the other squad that was pinning down the enemy. We did not know the ledge had broken. We did not know our brothers needed our help. We heard the firing shots but kept on waiting for orders from Sanjiv Saab. Leaving well laid positions were direct defiance of his orders.'

While clearing the area and taking precautionary headshots of the terrorists in the morning, each squad member cried. They were tears of pride but also of deep sadness. There is a myth among The Fourth of North that many of them shared with me, saying, 'We have been a part of hundreds of operations and killed hundreds of militants—but our casualty rate has always been so low that we always told ourselves *4 Para ke bando ko goli nahi lagati* [the men of 4 Para will never get hit by bullets). The myth was broken that day.'

Havildar Devendra Singh was still conscious and was evacuated immediately, but he had lost so much blood that he could not be saved. He went away gloriously but left behind his wife Vinita. Young, pretty Vinita has lost all her sheen. It has taken her a long time to believe that her husband has left her alone in the world. The world feels different now to her. The pride is there but the pain is visible too. How painful it must be to lose your husband as a young wife! How would it feel to deal with the shattered dreams of raising children together, building that house and so on? I had no words to console her, so we just hugged for a long time. Army wife to another wife, we were 'sisters-in-arms'.

Almost an entire squad had been annihilated, but their sacrifice was not in vain. They had not only completed a four-day-long operation in which various Army units were involved, but they had also added a glorious new chapter to the history of the Indian Army. Their operation would become a case study taught in military training centres. The squad affirmed the ethos and spirit of the Indian Army to the people of India. The deep ties of brotherly comradeship that the Indian Army is known for was displayed most magnificently by these five men who made the supreme sacrifice.

When news of Operation Rangdori Baihk was made public the following day, on 6 April 2020, along with accompanying pictures, the world was grappling with the reality of the newly arrived Covid-19 pandemic, and the nation was under complete lockdown. The unit also faced tremendous challenges to deport the bodies to their hometowns. No larger-than-life ceremonies could be held for them due to Covid restrictions but still, the whole nation mourned this sacrifice. This operation gave hope and courage to the people of India, who were struggling with the deadly pandemic. These five operatives from the 4 Para (SF) became household names.

Later, Subedar Sanjiv Kumar was posthumously awarded the Kirti Chakra for displaying outstanding leadership, raw courage and utmost gallantry beyond the call of duty. It is one of the three Kirti Chakras that 4 Para (SF) has. And Havildar Devendra Singh, Paratrooper Bal Krishan, Paratrooper Amit Kumar and Paratrooper Chattrapal Singh received the Sena Medal posthumously.

When I left the unit office area, I turned around one last time and found their portraits shining gloriously on the 'Roll of Honour' board of 4 Para (SF) placed at the main entrance. Their pictures were placed one after the other, together as one squad. They have been successful in immortalizing their camaraderie forever.

I also found this bonhomie, deep compassion and empathy amongst women behind the brave men of 4 Para (SF). I was invited to give a lecture at a welfare programme during their celebrations for the battalion's raising day. Operatives posted abroad, families, women and children came from far-flung villages and cities. Since men mostly live at forward posts, SF families settle all across the country. Despite a beautiful programme, with music and dance, the atmosphere suddenly got heavy when the Commanding Officer's wife, started felicitating the 'Veer Naris and Veer Matas' of 4 Para (SF). Each one of them would come—mothers and wives—hug her and cry their hearts out. Needless to say, each one of us sitting in the audience were crying too. When the first lady of the 4 Para (SF) requested me to give a lecture on *The Force Behind the Forces*, I had to cancel two of my confirmed contractual events to be there. I am glad I made that decision. One needed to be present there to witness the intensity of emotions of this close-knit group of women, whose husbands or sons have sacrificed their lives for the nation. Though I also met newly wedded brides glowing in the company of their husbands, who were dishevelled and in long beards and had come down from the forward locations at LAC and LoC just to celebrate their beloved unit's diamond jubilee. All the ladies that I met there had stories of immense sacrifices and the choices they made so their husbands could serve without any tension. Their lives and challenges invoke a lot of respect for the Army wives who work selflessly even without a share in the medal their husbands earn.

The people of India must know and remember this story. Many are unaware of the true ethos of the Indian Army. Heroism is not always about winning. This story highlights the fact that our soldiers and their families depend heavily on each other. Their comradeship and brotherhood are what have kept the Indian Army standing tall as the world's finest army. I hope people of India never forget who pays the cost of their freedom and dreams. The least we can do is to hold gratitude. Jai Hind!

This story is based on interviews conducted with the serving soldiers and officers of the immortals of 4 Para (SF). They are all serving SF operatives and want to remain anonymous. I also met mothers and wives of these operatives who have made supreme sacrifices in the line of duty. The women left behind are heartbroken but they are trying to move forward with their lives. The memories spent with the beloved are the sole source of their smiles. This story is a tribute to all the band of brothers serving under the most adverse situations and the families they have left behind. It was important to tell to let the people of India know that freedom is never free.

Parachute Regiment Insignia

Regimental Badge (Para Crest)

The badge has a canopy and rigging line of a parachute signifying the Regiment's airborne role. The wings of both side denote flight and have a direct link with the aircraft from where the paratroopers jump out. On the rigging lines is superimposed a bayonet implying that the regiment is an infantry arm. The scrolls on both side are decorative motifs.

Formation Sign

Shatrujeet, adopted from Hindu mythology, depicts the puranic king of the same name and his horse Kuvalya, who were inseparable and thus enjoined as one being. The figure holds a bow, taut with an arrow about to be delivered, riding his horse and using a divinely gifted arrow. The king was said to have slain a mighty demon. 50 (I)Para brigade uses the emblem on their dress, flag and stationery.

Balidan (Sacrifice) Badge

Awarded to personnel after they have spent one year with a PARA (SF) battalion or six months if the battalion is involved in active duty. Col Megh Singh first proposed the adoption of the 'Balidan' insignia. It portrays a dagger with wings and a motto 'Balidan' emblazoned on it. The badge signifies the unconventional warfare missions of the Special forces.

Special Forces Flashes

Worn on the arm, the flashes are awarded to personnel of Para (SF) units after undergoing a gruelling probation.

Diving badge

It depicts a diver fully geared with combat loads. The outstretched arms indicate that he is navigating under water while the helmet and the fins draw attention to the great depths at which divers operate, and that they strive to move ahead despite all odds. The Ashoka emblem symbolizes the sovereign state of India.

Wings

The paratrooper's wing is made of cloth and is worn on the right breast pocket above the nameplate. It is awarded after successful completion of the Parachute Basic Course.

Acknowledgements

I would like to extend my biggest thanks to the Commanding Officer of a Gorkha Battalion, who, during a dinner at my home, had casually remarked, 'Ma'am if you are planning to write on SF forces, what are you doing here and not hitting the jungles or mountains? Once you sit down with those guys, listen to their stories, *aapke jazbat badal jayenge, alfaz badal jayenge.*' And it changed the course of my research. This book has come into being due to the exceptional support, help and guidance of SF units. I am indebted to the senior officers, commanding officers, officers-in-arms, JCOs and jawans who supported me in my journey to immortalize the stories of the bravehearts.

I would like to thank the mothers, fathers, wives and children of the SF operatives, who not only answered thousands of my questions but also opened the doors of their homes and hearts for me. My thanks also to the friends and mentors of these SF operatives for sparing their time for me.

My gratitude to the Additional Directorate General of Public Information. Be it about accessing documents, applications, other correspondence or requiring approvals from various authorities and departments, I received complete support from ADGPI. I would also like to extend my thanks to the Indian Army fraternity, to every man and woman who is part of it. I want to tell each one of you that I feel exceptionally privileged to share space with you as an army wife. I hope I have made all of you proud in my little capacity.

My thanks to Kanishka Gupta and Narayani Basu for the effort they have put in the book. To Swati Chopra for commissioning the book, believing in it and extending the best of resources. I am thankful to the designers, editors and everyone at HarperCollins India who worked day and night to make *Balidan* better.

I have immense gratitude for my parents and my brothers. It is with your love and support that I am where I am today and doing what I love to do. Special mention to my husband—you are a dream come true. Thank you for looking after the kids, who are usually on the mission to not let Mommy write. If you think soldiers only fight on the battleground, you are mistaken. Some challenges are also thrown by their children and working wives.

I would also like to thank my readers for believing in my work and supporting me fiercely. I hope you understand that I have tried to do my bit by bringing forward meaningful stories to inspire people. I wanted to tell the stories of the real superheroes of India to our children. Now that the book belongs to you, please ensure that every Indian reads it and take pride in the existences of such Kohinoors. It's unfortunate that many of the gems made supreme sacrifices. They were supposed to stay amongst us and cherish their mention in our hearts and gratitude. I apologize for any mistakes that I may have made in this book. It was an arduous task and I have tried my best.

In the end, my biggest thanks to Mata Rani. I was not capable of writing something as extraordinary as a book on the SF—a book that has six extensively researched, authentic and inspiring stories. Only you and I will know all that it took to make this book happen. I hope it serves your purpose in the most glorious manner, Divine Mother.

For our nation, for the people!

Jai Hind!

About the Author

Swapnil Pandey's extensively researched books are aimed at creating awareness about the work and lives of Indian Army personnel and their families. These include *The Force Behind the Forces*, *Love Story of a Commando* and *Soldier's Girl*. She is an alumnus of Birla Institute of Technology, Mesra. She has worked with organizations like Wipro and HDFC, and taught at Lovely Professional University and the Army Public School. She writes a popular blog as well as articles for magazines and newspapers, and appears on national TV as a panellist. As a prominent voice from the army fraternity, she has been felicitated by organizations like ISRO. She can be contacted at teamgirlandworld@gmail.com, on Twitter at @swapy6, on Facebook at 'Author Swapnil Pandey', and on Instagram at @swapnil_pandey_author.

HarperCollins *Publishers* India

At HarperCollins India, we believe in telling the best stories and finding the widest readership for our books in every format possible. We started publishing in 1992; a great deal has changed since then, but what has remained constant is the passion with which our authors write their books, the love with which readers receive them, and the sheer joy and excitement that we as publishers feel in being a part of the publishing process.

Over the years, we've had the pleasure of publishing some of the finest writing from the subcontinent and around the world, including several award-winning titles and some of the biggest bestsellers in India's publishing history. But nothing has meant more to us than the fact that millions of people have read the books we published, and that somewhere, a book of ours might have made a difference.

As we look to the future, we go back to that one word—a word which has been a driving force for us all these years.

Read.

Harper
Collins

HARPER
FICTION

HARPER
NON-FICTION

HARPER
BUSINESS

HARPERCOLLINS
CHILDREN'S BOOKS

HARPER
DESIGN

Harper
Sport

HARPER
PERENNIAL

HARPER
VANTAGE

हार्पर
हिन्दी

BOOKTOPUS